D1265774

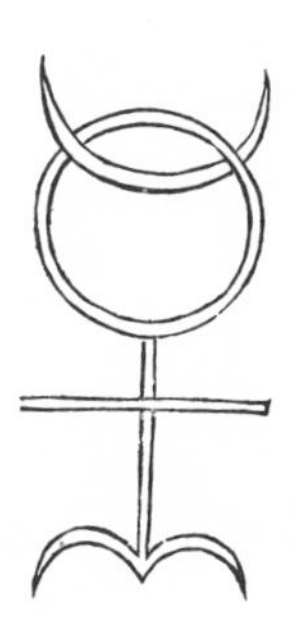

PRIMARY SOURCES FROM

THE SCIENTIFIC REVOLUTION

General Editor, Allen G. Debus

Director, The Morris Fishbein Center
for the Study of the History
of Science and Medicine
The University of Chicago

D.R JOHN DEE.

THE MATHEMATICALL PRAEFACE

to the

ELEMENTS OF GEOMETRIE

of Euclid of Megara

(1570)

with an Introduction by

ALLEN G. DEBUS

SCIENCE HISTORY PUBLICATIONS
NEW YORK

First published in the United States by
Science History Publications
a division of
Neale Watson Academic Publications, Inc.
156 Fifth Avenue, New York 10010

First Edition 1975
Designed and manufactured
in the U.S.A.

Library of Congress Cataloging in Publication Data

Dee, John, 1527-1608.
The mathematicall praeface to the Elements of geometrie of Euclid of Megara (1570)

(Primary sources from the scientific revolution)
1. Mathematics--Early works to 1800. 2. Philosophy --Early works to 1800. 3. Geometry--Early works to 1800. I. Title.
QA33.D4 516'.2 74-31377
ISBN 0-88202-020-X

INTRODUCTION

HERE are few figures of Renaissance science whose work has been the subject of more divergent interpretations than John Dee. To some he represents one of the most illustrious of the sixteenth century occultists, a man whose primary interests were magic, astrology and alchemy. To others he stands as a key figure in the development of modern science. Yet, even here we find little agreement. Professor E. G. R. Taylor would have us forget Dee's mysticism. Rather, she has described his astronomical and navigational instruments while referring with approval to the contemporary references by Dr. Richard Foster and Tycho Brahe. The former spoke of Dee "as a very Atlas bearing upon his shoulders the sole weight of the revival in England of the mathematical arts." The latter was to address Dee and Thomas Digges as "most noble, excellent and learned" mathematicians.[1] Dr. Frances A. Yates would not disagree with Professor Taylor in regard to the high order of Dee's significance, but for her Dee's primary role was to serve as a major link between sixteenth century Hermeticism and the "Rosicrucian" authors of the seventeenth century.[2] She has pictured this influence as an essential ingredient in the development of modern science. Thus Taylor would call Dee a key figure in the Scientific Revolution in spite of his mysticism—and Yates would do the same because of it.

In their judgments of Dee both Taylor and Yates most frequently cite the "Mathematicall Praeface" prepared for the first English translation of Euclid's *Geometry* made by Henry Billingsley (1570). Here is to be found a classification of the mathematical arts—along with descriptions of each one. It is an early account of this kind—and one that was frequently referred to in the final decades of the sixteenth century. And yet, although the "Mathematicall Praeface" is cited by most historians of science, it has only been reprinted in its entirety twice—first in 1651 and once again ten years later. The brevity of this text combined with the frequently repeated claims of its importance would indicate that a new edition is long overdue. Accordingly, the present reprint—from the first edition of 1570—needs no apology.

The Life of John Dee

John Dee was born on the thirteenth of July, 1527, the son of a London mercer, Roland Dee, and his wife, Johanna Wilde Dee.[3] At the age of fifteen Dee entered St. John's College, Cambridge where his talent in mathematics was discovered by the noted humanist and student of Greek mathematics, Sir John Cheke. After taking his B.A. (1545), Dee became a fellow of St. John's and later (1547) a foundation fellow of Trinity College. It was there that he contributed a mechanical flying scarab to a College production of the *Pax* by Aristophanes —"whereat was great wondering, and many vaine reportes spread abroad of the meanes how that was effected."[4] It is evident that Dee had a real interest in the Hellenistic tradition of mechanical marvels even at this early date.

In 1547 Dee made the first of several journeys to the continent. Here he sought out numerous scholars, but "chiefly mathematicians." Among these were Gemma Frisius and Gerard Mercator. They encouraged him in his interest in geography and mapmaking and on his return home he carried "the first astronomer's staff or brass, that was made of Gemma Frisius' devising, the two great globes of Gerardus Mercator's making, and the astronomer's ring of brass, as Gemma Frisius had newly framed it."[5] Entries in his diary note that Dee was beginning "to make observations (very many to the houre and minute) of the heavenly influences and operations actuall in this elementall portion of the world. Of which sort I made some thousands in the yeares then following . . ."[6] This quotation indicates that Dee's interests were directed as much toward astrological as astronomical studies. And although he was willing to use the heliocentric system of Copernicus for calculations, it would be erroneous to dismiss his belief in astrology as being lightly held.

After taking the degree of Master of Arts (1548), Dee left again for the continent—this time for a three year span. The first two of these years he was a student at Louvain. His reputation was already considerable and many noblemen came to visit him from all parts of Europe.[7] It was also at this time that he began the practice of private instruction—a custom he continued for many years. In mid-summer, 1550, Dee proceeded on to Paris where he added immeasurably to his renown.

> I did undertake to read freely and publiquely Euclide's Elements Geometricall, *Mathematicè, Physiquè, et Pythagoricè*. . . . My auditory in Rhemes Colledge was so great, and the most part elder than my selfe, that the mathematicall schooles could not hold them; for many were faine, without the schooles at the windowes, to be auditors and spectators, as they best could helpe themselves thereto. I did also dictate upon every proposition, beside the first exposition. And by the first foure principall definitions representing to the eyes (which by imagination onely are exactly to be conceived) a greater wonder arose among the beholders, than of my Aristophanes *Scarabaeus* mounting up to the top of Trinity-hall in Cambridge *ut supra*.[8]

Still a young man, John Dee was already internationally famous. He rejected an offer "to be one of the French king's mathematicall readers" and in a later year (1592) was to recall that "I might have served five Christian Emperor's; namely, Charles the Fifth, Ferdinand, Maximilian, this Rodulph, and this present Moschovite."[9] Choosing to return to England, Dee was presented to the King by William Cecil and after that time rose rapidly in Court circles. He served as an astrologer to Edward VI, Mary and Elizabeth—and for this was accused by his enemies of practicing black magic. Imprisoned (1555) under the charge of treason, he was acquitted,[10] but this was not to be the last time he was to suffer because of his work. Not many years later the Calvinist, John Foxe, referred to Dee as "the great Conjurer" in his widely read *Actes and Monuments* (1563)[11] and this charge was to plague him intermittently throughout the remainder of his life. Nevertheless, like most of those at Court, Dee found it possible to adjust to the religious upheavals of the period. Honored by the Protestant, Edward, he was consulted by the Catholic, Mary—while he was to entertain and discuss his books with Elizabeth at his home at Mortlake.

Dee spent most of the years 1551–1583 in England. His life there was pleasant and conducive to study. His sovereigns supported him in a fashion he found comfortable and it is possible to follow his interests through his numerous writings and through the several manuscript copies of his extensive library. This collection was built over the course of these decades and it was a possession of great pride to Dee. He had been genuinely alarmed at the dispersal of the monastic libraries and had urged Queen Mary to preserve whatever might still be saved from this wholesale destruction of the learning of the past.[12] When his proposal came to naught, he himself began to build a major library that was to include ancient, medieval and contemporary authors. Dee was to write that it eventually comprised some three thousand printed books and an additional thousand manuscripts. Commenting on the latter, M. R. James has noted that the collection was a

> first class repository of medieval science books excluding medicine. Alchemy, astrology, astronomy, physics, geometry, optics, mathematics are all very copiously represented . . . There is very little theology and no ancient poetry . . . History, British and English, is perhaps the subject best represented next to Natural Science.[13]

And although more medicine was to be found in the printed collection, in general one could say that the same breakdown would hold true for the published works.[14]

We may follow Dee's interests also from his bibliography[15]—and here again we may note the same emphases seen by James. Dee's earliest writings (1547, 1548) dealt with logic, but by the early fifties there was a dominant interest in mathematical and astronomical-astrological themes. For the years 1550–1551 Dee referred to his tracts *De uso Globi Coelestis* and a description of the cosmos prepared for Edward VI. A copy of his Parisian lectures on Euclid dates from the same period and this was most likely employed to good use in his notes on Euclid and the "Mathematicall Praeface" twenty years later. Studies of the tides, perspective, and burning mirrors were all mathematical subjects completed in the fifties while an apology for Roger Bacon defending him against the accusation of having used demons (1557) surely reflected his own recent problems.

Dee has rightly been praised for his introduction to the revision of Reinhold's Prutenic tables prepared by John Feild (1556).[16] It was Dee who had encouraged his friend to base his recalculation of these tables on the Copernican theory. In his earlier *De Nubium, Solis, Lunae, ac reliquorum Planetarum* he had specifically applied himself to the calculation of the distances of the planets from the center of the earth while the Queen and the Privy Councilors had been responsible for his plan to reform the Calendar (1582).[17] No less important was Dee's tract on the new star of 1572, the *Parallaticae Commentationis Praxeosque Nucleus quidam* which included trigonometric theorems for determining stellar parallax.[18]

Still, Dee's work in this field had broader implications than would be encompassed by theoretical and mathematical astronomy. His early contact with Gemma Frisius and Mercator—his persistent collecting of astronomical instruments—were closely connected with his interest in navigation and geography. He wrote a number of manuscripts on these subjects which for the most part remained unpublished. For three decades he was to serve as an advisor to English mathematicians and navigators. As a consultant he visited Muscovy House when Martin Frobisher was about to depart on his first voyage in search of the Northwest Passage. There he lectured the chief pilot and the mariners on the rules of geometry and instructed them in the use of the navigating instruments that were to be used on the journey.[19] His knowledge was so esteemed by Sir Humphrey Gilbert that he entered into an agreement that would have made him the owner of most of what is now Canada.[20]

Dee's desire to improve astrology is evident in his *Propaedeumata aphoristica . . . de praestantioribus quibusdam naturae virtutibus* (1558). This is composed of one hundred twenty aphorisms which he had been prompted to prepare after receiving a letter from Mercator on current disputes among astrologers. In his aphorisms Dee assumed that the subject could be approached mathematically and on this basis he proceeded to investigate the nature of astral influences and the theory of signatures. The former, he felt, were both natural and predicatable, thus providing a fundamental mechanism for causality. Dee was convinced that his work provided a new and certain foundation for astrology—a science that might now be based on mathematics in a fashion not previously possible.[21]

No less fundamental than astrology was alchemy or "astronomia inferior," the subject of his *Monas Hieroglyphica* (1564).[22] Here, too, Dee was convinced that a new approach was necessary in order to realize the true potential of alchemy as a key to nature. Arguing that "Pythagorean" symbols have an underlying clarity and strength "almost mathematical,"[23] Dee proceeded to construct a figure based on points, straight lines and circles through a series of "theorems." The finished product closely resembled the alchemical figure for mercury. Earlier, this Monad had figured prominently on the title page of the *Propaedeumata aphoristica* and it was again to be displayed on the first page of the "Mathematicall Praeface." For Dee this symbol might be understood as "the rebuilder and restorer of all astronomy."[24] In the process of its construction the scholar had himself produced the sun, the moon and the earth.[25] In effect he had—in symbolic form—retraced the first stages of the Creation. From this point the mysteries of the elements were developed. Throughout Dee digressed to offer numerical and kabbalistic analyses—and the latter part of the slim volume would appear to be a veiled representation of the alchemical process itself.[26] Proper understanding of the Monad was promised to reveal the true mysteries of the physical world since the symbol was alleged to include the hidden secrets of both celestial and terrestrial astronomy.[27] However, Dee's object was also closely allied to that of Renaissance magic in a broader sense. The mysteries of the Monad would enable the artificer to produce wonderful effects in engineering, music, optics, the manipulation of weights, and hydraulics.[28] It would also unite the different parts of Cabala—and make easily available all art and medicine.[29]

Dee's spiritual emphasis in the *Monas Hieroglyphica* was closely in tune with the mystical mathematics favored by those who sought truth in a Pythagorean tradition. Here he had sought a key to Creation in mysticism, natural magic and numerical analysis. In contrast, more conventional mathematical proofs, chemical laboratory techniques and practical medical applications were of relatively little interest. The *Monas Hieroglyphica* gained Dee considerable attention both in England and abroad—and it was a work he considered to be of basic importance. He was later to note on the fourteenth of June 1564 that

> After my retorne from the Emperor's court, her Majestie very graciously vouchsafed to account herselfe my schollar in my booke, written to the Emperor Maximilian, intituled, *Monas Hieroglyphica;* and said, whereas I had prefixed in the forefront of the book: *Qui non intelligit, aut taceat, aut discat:* if I would disclose unto her the secretes of that booke, she would *et discere et facere;* whereupon her Majestie had a little perusin of the same with me, and then in most heroicall and princely wise did comfort me and encourage me in my studies philosophicall and mathematicall, &c.[30]

The *Monas Hieroglyphica* was a theoretical treatise—one that accurately reflected the strong representation of alchemical books and manuscripts in Dee's library. Yet this collection and the book show Dee to have been an atypical chemist of this period. Although he owned many works by Paracelsus and even prepared his own translation of one of them into French, it would appear that he had very little interest in the alchemical-chemical medicine that characterized this school and dominated late sixteenth century chemistry. Rather, as one follows the entries in his diary and his other autobiographical works one is struck by the ever increasing interest in traditional alchemical experiments. His house at Mortlake housed three chemical laboratories equipped with expensive equipment.[31] This may well have been the largest chemical laboratory in England at that time.

In March of 1582 John Dee was visited at Mortlake by Edward Kelley,[32] a fellow alchemist who was to make himself indispensable by acting as a medium in experiments with a crystal. And, already deeply involved in these investigations,

Dee was to meet the Polish Prince Albertus Laski who had similar interests early in the following year.[33] Within a few months Laski invited Dee, Kelley and their families to Poland. Shortly after their departure (21 September 1583)[34] a mob broke into the house at Mortlake dispersing the collection of books, manuscripts and instruments and destroying the laboratories.[35]

On the Continent Dee travelled widely, but spent most of his time in Poland, at the court of the Emperor in Prague and at Trebona, a castle owned by Count Rosenberg of Bohemia. Dee and Kelley continued their mystical studies and alchemical experiments until 1589 when the former returned to England.[36] The relationship with Kelley had been a stormy one and his medium was to remain behind where he was first knighted, but eventually was to die in an abortive escape from prison (1595).

Dee's final years in England were ones of disappointment. Never again was he to regain the extent of royal patronage he had enjoyed prior to 1583. His *Compendious Rehearsall* (1592) and *Discourse Apologeticall* (1595) were written to remind Elizabeth of his long years of service and of his devotion to scholarship. Similar appeals were penned to King James and to Parliament in the first years of the new reign. In fact little was accomplished. The new King was an ardent opponent of witchcraft and there is little doubt that many thought Dee's credentials were open to question on this score. He was forced to relinquish his one remaining position as Warden of Christ's College, Manchester and—following the death of his third wife—he lived in poverty until his death in 1608 attended only by his daughter, Katherine.

The Background to the "Mathematicall Praeface"

Perhaps the most interesting single work written by John Dee is his "very fruitful" . . . "Mathematicall Praeface" prepared for the translation of Euclid's *Elements of Geometrie* made by Sir Henry Billingsley. Even a cursory reading of this introductory piece will reveal that any simple definition of mathematics would be insufficient to encompass Dee's approach to his subject. As the man was attracted to a mathematical spectrum that ranged from the study of navigation and mechanics to mysticism, so too his "Praeface" reflected the study of this subject on all levels. For this reason some background relating to the late medieval and Renaissance mathematical themes familiar to John Dee is necessary for any discussion of the "Mathematicall Praeface."[37]

The Renaissance was a period when the traditional scholasticism of the universities was under widespread attack. Many scholars—seeking a new approach to knowledge—turned to the study of the neo-Platonic and Hermetic writings—sources which were thought to have strong and valid connections with Christianity. Some were to find truth in an alchemical cosmology, but, for nearly all, this search resulted in a new emphasis on mathematics as a key to the universe. Thus for Nicholas Cusanus (1401–1464) both philosophical truth and Holy Writ lead to the same conclusion, that a knowledge of mundane and divine truth may be obtained through a knowledge of numbers. In *Wisdom* 11:7 (Douay version 11:21) we read that God created "all things in number, weight and measure."[38] The Creation itself may then be interpreted as a mathematical process since arithmetic, geometry, music and astronomy are the "same sciences of God employed when He made the world."[39] Thus, by a mathematical study of nature we enrich ourselves through a deeper understanding of the Creation and the Creator.

Yet, for Cusanus more than Biblical truth leads us to this conclusion. Philosophers agree that the visible universe is a faithful reflection of the invisible one and we may surely expect to rise to a firm knowledge of our Creator if we properly study the creatures and objects of our world.[40] But a study of the multiple beings in nature can only lead to our belief in an underlying unity pervading all things. We don't know why, but

> we know for a fact that all things stand in some sort of relation to one another, that, in virtue of this inter-relation, all the individuals constitute one universe and that in the one Absolute the multiplicity of beings is unity itself.[41]

Cusanus in this way was led to believe that direct observations in nature were a valid approach to truth. Still, no amount of experimental knowledge could give us complete understanding because of the imperfection of our own world. When possible we must begin our search in the perfect world of ideas where we may rely on mathematics which starts with the finite, but allows us to progress confidently to the simple infinite. For Cusanus mathematics stood as the perfect example of abstract truth.

> That explains why philosophers so readily turned to mathematics for examples of the things which the intellect had to investigate; and none of the masters of old, when solving a difficulty used other than mathematical illustrations, so that Boethius, the most learned of the Romans, went so far as to say that knowledge of things divine was impossible without some knowledge of mathematics.[42]

Cusanus turned not only to Boethius, but also to the Pythagoreans, St. Augustine, Plato and Aristotle for examples of the importance of mathematics. It was especially in the neo-Platonic and Pythagorean traditions that the cosmological significance of mathematics was maintained. This could be illustrated through the work of a series of authors ranging from late antiquity through the middle ages. Cusanus could thus rightly call on the testimony of many earlier philosophers who had pointed to mathematics as a key to the Creation epic and therefore to nature as a whole.[43]

Because of the breadth of the subject matter covered by mathematics it was valid to question just how it might best be studied. All agreed that arithmetic had useful applications for businessmen and artisans, and on a higher level there was agreement that the beauty of higher mathematical proofs would prepare the mind for divine truths. Mysticism in mathematics was most closely identified with the neo-Platonic and neo-Pythagorean texts which were filled with accounts of angels, demons and magic. It is hardly surprising that such studies led early to distrust on the part of theologians. As early as the thirteenth century Roger Bacon had found it necessary to defend the true mathematics (which he derived from *mathesis* with a short middle syllable—and meaning knowledge) from the false mathematics or magic (derived from *mathesi* with a long middle syllable—meaning divination).[44] The revived interest in neo-Platonic thought in the fifteenth century made this a question of vital interest. No one denied the existence of evil and demonic magic, but surely all forces in the universe must be natural in their operation since they all derive from the omnipotent Creator. It was only the goal of the magician which might be questioned. Most natural magicians of the Renaissance answered their opponents by affirming that their task was not to prepare spells, charms and incantations. Rather, they explored the wonders of God's created universe so that they might better comprehend their Creator. Since this was the meaning of true magic, its practitioners could and did take pride in their accomplishments.

The equating of astrology, astronomy and mathematics is a persistent theme in Renaissance texts and one which is referred to by authors of widely differing views. Thus Pico della Mirandola insisted that the first two subjects were only parts of the third while it, in turn, was essential for further studies in theology.[45] For some, mathematics really meant iatromathematics, the use of astrology in medicine.[46] Leonard Digges opened a practical treatise on surveying with a defence of astronomy—or rather astrology—and the mathematical sciences.[47] Porta, in his *Natural Magick*, explained that magic is essentially the search for wisdom and seeks nothing else but the "survey of the whole course of nature."[48] In this legitimate and worthy search the Godly *magus* is told that he "must know the Mathematical Sciences, and especially Astrologie."[49] Similar cosmic views are to be found in Agrippa's *De Occulta Philosophia*. Here magic is depicted as the most perfect science, the consummation of philosophy.[50] A magician, however, can do nothing without mathematics since all things were made and continue to be ruled over by number, weight, measure, harmony, motion and light. These are mathematical studies which are necessary for natural philosophy. When they are mastered they may be put into practice with the mechanical arts to produce marvels and wonders.[51] Man himself might fruitfully be subjected to a numerical analysis. He is the most nearly perfect product of God's Creation and in him will be found the same composition according to weight, number and measure that we know exists in the macrocosm. Therefore, the proportions and harmonies we establish in the heavens and the earth must also be found by valid analogies in our own bodies.[52] For Agrippa, these truths might easily—and very properly—be related to kabbalistic studies.

Thus the Renaissance "mathematician" could think of himself primarily as a student of universal harmonies—a man for whom the investigation of mystical relationships of the microcosm and the macrocosm would seem to be on a higher plane than the direct calculations of the "common" arithmetician. Such a man might also be attracted to natural magic and to the duplication of those mechanical wonders described by ancient and medieval authors. Yet, at the same time, he would not deny that the search for God's plan of Creation as seen in his numbering, his weighing, and his measuring might be studied—if on a less lofty plane—through new observations in nature. Nor would he deny the purely practical value to be found in the application of mathematical studies to everyday affairs.

The "Mathematicall Praeface" (1570)

John Dee was particularly well suited to comment on Euclid. His training under Sir John Cheke at Cambridge had been followed by his public lectures on the *Elements* in Paris. As a classicist he critically studied the Billingsley translation and freely corrected imperfectly understood passages. In addition, he prepared annotations, new corollaries and alternate proofs when he saw fit to do so. But as he worked over the text in the late 1560s, Dee felt that he had to answer why this—or any—learned treatise should be translated into a common language. It was a question that was to be discussed frequently by sixteenth century translators. On this point Dee argued that although he knew that some would disagree, he was convinced that there really was no threat to the universities in the publication of Euclid's work in English. Similar translations into Italian, German, Spanish and French already existed and they had not harmed the universities in those countries. Rather, students would now be instructed more rapidly in mathematics and some, realizing that the universities are "the Storehouses & Threasory of all Sciences, and all Artes," would be attracted to these universities for advanced study.

> Besides this, how many a Common Artificer, is there, in these Realmes of England and Ireland, that dealeth with Numbers, Rule, & Cumpasse: Who, with their owne Skill and experience, already had, will be hable (by these good helpes and informations) to finde out, and deuise, new workes, straunge Engines, and Instrumentes: for sundry purposes in the Common Wealth? or for priuate pleasure? and for the better maintayning of their owne estate?[53]

Mathematics was a key to knowledge and all should have it available to them—the unlatined as well as those at the universities.

But more than a vindication of the need to translate into the vulgate seemed necessary. Accordingly Dee planned a lengthy preface that would indicate the essential nature of mathematics to those who were about to work their way through Euclid. The resultant work was to become far better known than the translation—and it was this upon which Dee's reputation as a mathematician was largely based for the next century. Dee himself was very proud of the "Mathematicall Praeface" and he was later to comment that in it he had described "many Arts, of me, wholy invented (by name, definition, propriety and use,) more then either the Graecian, or Roman Mathematiciens, have left to our knowledge."[54] But if Dee was to consider this one of his major works, there is evidence to indicate that the "Mathematicall Praeface" was prepared under considerable pressure. More than once he wrote of having delayed an impatient printer. A series of dates tends to confirm this urgency. The publication date of the translation itself is given as February third, 1570. Dee dated his "Praeface" "At Mortlake. Anno 1570. Februarij 9"—while the folding "Groundplat of my MATHEMATICALL Praeface" (evidently the last part of the volume to be printed) is dated the twenty-fifth of February.[55]

Although Dee was at all times faced with pressure from the printer, the resultant work remains one of the most interesting Renaissance discussions of mathematics and its applications. He began the "Mathematicall Praeface" by noting that the intent of the work would be to describe "that mighty, most pleasaunt, and frutefull *Mathematicall Tree*, with its chief armes and second (grifted) braunches: Both, what euery one is and also, what commodity, in generall, is to be looked for, as well griff as stocke."[56] However, prior to proceeding to the various branches of the subject, Dee paused to point out the fundamental nature of numbers. Boethius had written that "*all thinges . . . do appeare to be Formed by the reason of Numbers. For this was the principall example or patterne in the minde of the Creator.*"[57] Pico had been no less definite in noting that "*by Numbers, a way is had, to the searchying out, and understandyng of euery thyng, hable to be knowen.*"[58] For Dee, God's "*Numbryng*, then, was his Creatyng of all things. And his Continuall *Numbryng*, of all thinges, is the Conseruation of them in being."[59] Therefore "the constant law of nũbers . . . is planted in thyngs Naturall and Supernaturall: and is prescribed to all Creatures, inviolably to be kept."[60] It stands to reason then that the study of numbers will lead us to a greater knowledge of all things. In this way

> we may both winde and draw our selues into the inward and deepe search and vew, of all creatures distinct vertues, natures, properties, and *Formes:* And also, farder, arise, clime, ascend, and mount vp (with Speculatiue winges) in spirit, to behold in the Glas of Creation, the *Forme* of *Formes*, the *Exemplar Number* of all thinges *Numerable:* both visible and inuisible: mortall and immortall, Corporall and Spirituall.[61]

Basic to mathematics is the study of *Arithmetic*, "the *Science that demonstrateth the properties, of Numbers, and all operatiõs, in numbers to be performed.*"[62] This is a divine subject which may properly be ranked second only to Theology.[63] Although it is essential to the merchant, it is no less valuable to others. Thus the physician may apply the arithmetical Art of Graduation to the mixing of medicines which require both the proper degrees of heat and moisture to combat illness.[64] No less important are arithmetic rules for the tactics of the military commander while an understanding of proportions is essential to law and the assurance of justice.[65]

The other basic division of mathematics is *Geometry*, the science of magnitude. This originated with the measurement of land in Egypt, but although it has its practical applications, geometry leads also to far more sublime knowledge. It prepares us

> to conceiue, discourse, and conclude of things, Intellectual, Spirituall, aeternall, and such as concerne our Blisse euerlasting: which otherwise (without Speciall priuiledge of Illumination, or Reuelation frõ heauen) No mortall mans wyt (naturally) is hable to reach unto, or to Compasse.[66]

For this reason "I would gladly shake of, the earthly name, of Geometrie."[67] Far more suitable would be "*Megethologia:* not creping on ground, and dasseling the eye, with pole perche, rod or lyne: but liftyng the hart aboue the heauens, by inuisible lines, and immortall beames: meteth with the reflexions, of the light incomprehensible: and so procureth Ioye, and perfection vnspeakable."[68]

From this point Dee proceeded to describe a large number of disciplines and subjects based on mathematics which he divided into the two divisions, "Art" and "Art Mathematicall Derivative." The first category comprised methodologically complete doctrines, "the knowledge whereof, to humaine state is necessarye." Here were included the use of arithmetic and geometry for purely mathematical considerations and their application to supernatural and theological speculation. However, the emphasis of the "Praeface" is clearly on the more practical derivatives. An "Art Mathematicall Derivative" is that

> which by Mathematicall demonstratiue Method, in Nũbers, or Magnitudes, ordreth, and confirmeth his doctrine,, as much & as perfectly, as the matter subiect will admit. And for that I entend to use the name and propertie of a *Mechanicien*, otherwise, then (hitherto) it hath ben used, I thinke it good, (for distinction sake) to giue you also a brief description, what I mean therby. A Mechanicien, or a Mechanicall workman is he, whose skill is, without knowledge or Mathematicall demonstration, perfectly to worke and finishe any sensible worke, by the Mathematicien principall or deriuatiue, demonstrated or demonstrable. Full wel I know, that he wich inuented, or maketh these demonstrations, is generally called *A speculatiue Mechanicien:* which differeth nothyng from a *Mechanicall Mathematicien.*[69]

Turning then from the "art" of geometry it seemed possible to enumerate a number of "derivatives" which form part of "vulgar" geometry. Generally these are connected with measurement: of land, the depths of wells, the volume of casks and other determinations of this sort. The astronomer may similarly use vulgar geometry to measure the height of "blasing stars, and of the Mone" as well as their distances and solidities.[70] Major disciplines here include *Geographie*, and its derivative, *Chorographie*, *Hydrographie* (the study of the Oceans and sea coasts) and *Stratarithmetrie* (the application of geometry to warfare).[71] In addition to these "Artificiall Feates" dependent upon vulgar geometry there are a number of "*Methodicall Artes*" based upon the science of Geometry: *Perspectiue*, *Astronomie*, *Musike*, *Cosmographie*, *Astrologie*, *Statike*, *Anthropographie*, *Trochilike*, *Helicosophie*, *Pneumatithmie*, *Menadrie*, *Hypogeiodie*, *Hydragogie*, *Horometrie*, *Zographie*, *Architecture*, *Nauigation*, *Thaumaturgike* and *Archemastrie*.

Natural Philosophy, Dee affirmed, could not be understood without a knowledge of *Perspectiue*. This subject demonstrates "the manner, the properties, of all Radiations Direct, Broken, and reflected."[72] We must correctly interpret our observations—and this could be done only through the rules of this science. Thus, although on occasion two or three suns may be observed, this does not mean that there is more than one. And—if we learn the secrets of this science correctly—we may be able to produce optical marvels ourselves.

Astronomie is also "an Arte Mathematicall" and closely associated with perspective since it "demonstrateth the distance, magnitudes, and all natural motions, appearances, and passions propre to the Planets and fixed Sterres: for any time past, present and to come . . ."[73] It is only through careful and tedious observations and calculations that we may obtain the facts which allow us to determine "the Course of Times, dayes, yeares, and Ages." Still, astronomy has a higher aim through which we may interpret "Sacred Prophesies" and "high Mysticall Solemnities."[74] A sister science is *Musike* which goes beyond the pleasures of common songs and ballads to investigate the harmonies inherent in the universe. Similarly, *Cosmographie* "is the whole and perfect description of the heauenly, and also elementall parte of the world, and their homologall application, and mutuall collation necessarie."[75] This is the study of universal correspondences present in the macrocosm and the microcosm.

Astrologie for Dee is also a mathematical art, but one which is distinct from astronomy since it endeavors to demonstrate "the operations and effectes, of the naturall beames, of light, and secrete influence: of the Sterres and Planets: in euery element and elementall body: at all times, in any Horizon assigned."[76] Aristotle is a prime source of evidence in support of the action of the heavenly bodies on man and for Dee there could be no doubt "that mans body, and all other Elementall bodies are altered, disposed, ordred, pleasured, and displeasured, by the Influentiall working of the *Sunne*, *Mone*, and the other Starres and Planets."[77] But this art has been plagued by blind and unlearned followers who have only brought discredit on it. For truth the reader is told to read Dee's *Propaedeumata aphoristica* in which "I haue Mathematically furnished vp the whole method."[78]

Considerably removed from the preceding disciplines is *Statike* "which demonstrateth the causes of heauynes, and lightnes of all thynges: and of motions and properties, to heauynes and lightnes, belonging."[79] This subject is studied through the balance, one additional key to the Creation.

> Thou onely, knowest all thinges precisely (O God) who hast made weight and Balance, thy Iudgement: who has created all things in *Number, Waight, and Measure:* and hast wayed the mountaines and hils in a Balance: who hast peysed in thy hand, both Heauen and earth.[80]

The student of Statics was referred primarily to the work of Archimedes, and Dee himself translated six fundamental propositions (*2*, *3*, *5*, *6* and 7 from Book I; *1* from Book II) from the treatise *On Floating Bodies.* However, he was well aware that more recent work was in print which might correct errors which were to be found in the commonly held (Aristotelian) views on falling bodies. Through the study of this art

> great Errors may be reformed, in Opinion of the Naturall Motion of thinges, Light and Heauy. Which errors, are in Naturall

> Philosophie (almost) of all mẽ allowed: to much trusting to Authority: and false Suppositions. As, Of any two bodyes, the heauyer, to moue downward faster then the lighter. This error, is not first by me, Noted: but by one *Iohn Baptist de Benedictis*. The chief of his propositions, is this: which seemeth a Paradox.
>
> If there be two bodyes of one forme, and of one kynde, aequall in quantitie or vnaequall, they will moue by aequall space, in aequall tyme: so that both theyr mouynges be in ayre, or both in water: or in any one Middle.[81]

This is an early and significant reference to the work of Giovanni Battista Benedetti (1530–1590) whose work was influential on the young Galileo. Beyond this Dee argued that a knowledge of statics leads to the correct understanding of weight proportions in all subjects ranging from gunnery to medicine. And again in the Archimedean tradition Dee found a deep fascination in problems related to cubature and quadrature. There was no time to discuss all of this in detail.

> Although, the Printer, hath looked for this Praeface, a day or two, yet could I not bring my pen from the paper, before I had giuen you comfortable warning, and brief instructions, of some of the Commodities, by *Statike*, hable to be reaped: In the rest, I will therefore, be as brief, as it is possible . . .[82]

Quite different from Statike is *Anthopographie*, "the description of the Number, Measure, Waight, figure, Situation, and colour of euery diuerse thing, conteyned in the perfect body of MAN: with certain knowledge of the Symmetrie, figure, waight, Characterization, and due locall motion, of any parcell of the sayd body, assigned: and of Nũbers, to the sayd parcell appertainyng."[83] This is analogous to astronomy (the mapping of the heavens), geography (the mapping of the earth) and cosmography (the matching of both). No less than in the heavens and the earth will we find universal harmonies expressed in the microcosm. How could it be otherwise since man "participateth with Spirites, and Angels: and is made to the Image and similitude of God."[84] The knowledge of this science is to be determined through the study of subjects as diverse as anatomy, physiogonomy, chiromancy, metaposcopy and the rules of perspective.

A series of short descriptions follows to account for *Trochilike*, *Helicosophie*, *Pneumatithmie*, *Menadrie*, *Hypogeiodie*, *Hydragogie*, *Horometrie* and *Zographie*. The first is the study of circular motion—a subject which finds practical expression in the designing of mills. *Helicosophie* relates to spiral lines on conic surfaces and is useful for the architect and the designer of machines. The study of *Pneumatithmie* leads the scholar to an understanding of the relationships of water, air, smoke and fire. Here Dee was led to observe in traditional fashion that

> *Vacuum*, or *Emptines* is not in the world. And that, all Nature, abhorreth it so much: that, contrary to ordinary law, the Elementes will moue or stand. As, Water to ascend: rather then betwene him and Ayre, Space or place should be left, more then (naturally) that quãtitie of Ayre requireth, or can fill.[85]

The mathematical art of *Menadrie* comprises the study of the multiplication of forces through cranes and other lifting engines. This knowledge has a special importance for the designing and building of engines of warfare—best seen in the works Archimedes prepared for the defence of Syracuse against the Romans. *Hypogeiodie* will instruct the student how to construct and map tunnels and passageways underground—essential knowledge for the miner. *Hydragogie* permits the learned to pipe water to any desired spot so that fresh water may be led to cities through aqueducts or excess water led away from mines. The study of *Horometrie* demonstrates how "the precise vsuall demoninatiõ of time, may be known, for any place assigned."[86] Here a knowledge of sun dials and astronomical instruments is essential. Dee himself described here an aequinoctial dial he invented through which "(the Sunne shining) the Signe and Degree ascendent, may be knowen."[87] This subject is also one which properly leads to perpetual motion "which you shall (by furder search in waightier studyes) hereafter, vnderstand more of."[88] *Zographie*, also termed a mathematical art by Dee, is the study of painting and drawing.

Frances Yates has placed heavy emphasis on John Dee's section on *Architecture*.[89] Here Dee translated portions of the first book of the *De Architectura* by Vitruvius related to the training of an architect, subjects including painting, geometry, perspective, arithmetic, natural philosophy, music and astronomy. However, Dee continued,

> if you should, but take his boke in your hand, and slightly loke thorough it, you would say straight way: This is *Geometrie, Arithmetike, Astronomie, Musike, Anthropographie, Hydragogie, Horometrie, &c.* and (to cōclude) the Storehouse of all workmāship.[90]

And, Dee added, we may also consult a more recent author, Leone Battista Alberti, for confirmation of the fact that an architect must go beyond practical considerations.[91] In short, learned architects agree that a sound training in the mathematical arts is essential. Dee does reflect the Renaissance concept of an architect as a universal man—a theme portrayed in both Vitruvius and Alberti. And, although it may surely be argued whether Dee's interest in Vitruvius is a major basis for the English scientific revival, there can be little doubt that he was influential in the new English interest in the *De Architectura*.

More might have been written about Architecture, but once again Dee referred to the limited time he had been given for the completion of his work:

> Lyfe is short, and vncertaine: Tymes are perilouse: &c. And still the Printer awayting, for my pen staying: All these thinges, with farder matter of Ingratefulnes, giue me occasion to passe away, to the other Artes remainyng, with all spede possible.[92]

Turning to *Nauigation*, he pointed immediately to the many mathematical arts required by the master pilot. He must be able to use most of the instruments of the astronomer, be fully familiar with the variation of the compass, and know how to read hydrographical charts. An ability to calculate both longitude and latitude is essential. English pilots should be the most skillful in the world since God has "endued this Iland with, by reason of Situation, most commodious for *Nauigation*, to Places most *Famous* & *Riche*."[93] Dee singled out Sir Humphrey Gilbert as an example of the heights to which English excellence had reached in this art.

For Dee the word *Thaumaturgike* applied to "that Art Mathematicall, which giueth certaine order to make straunge workes, of the sense to be perceiued, and of men greatly to be wondred at."[94] Mathematicians skilled in this art are capable of producing all kinds of marvels through their ability to apply their mathematical knowledge. Thus, images may be made to appear where substance is absent, brasen heads may be made to speak and artificial beasts and birds made to move or fly.[95] This is in keeping with a long Hellenistic and medieval tradition—and reminiscent of Dee's youthful flying scarab. But while all these effects are accomplished through natural means, a "Digression Apologeticall" discloses just how sensitive Dee actually was to the charge of magic. In fact, those who produce these wonders seek to glorify their Creator. Why, then, "Shall that man, be (in hugger mugger) condemned, as a Companion of the Helhoundes, and a Caller, and Coniurer of wicked and damned Spirites?"[96] He himself had spent a fortune and nearly a quarter century on the accumulation of knowledge, only to be accused by some of conjury and sorcery.

> Well: Well. O (you such) my vnkinde Country men. O Brainsicke, Rashe, Spitefull, and Disdainfull Countrey men. Why oppresse you me, thus violently, with your slaundering of me: Contrary to Veritie: and contrary to your owne Consciences?[97]

Could he really be such a fool as "to forsake the light of heauenly Wisedome: and to lurke in the dungeon of the Prince of darknesse?"[98] On the contrary, I am "innocent, in hand and hart: for trespacing either against the lawe of God, or Man, in any my Studies of Exercises, Philosophicall, or Mathematicall."[99]

Dee closed his "Praeface" with *Archemastrie* which

> teacheth to bryng to actuall experience sensible, all worthy conclusions by all the Artes Mathematicall purposed, & by true Naturall Philosophie concluded: & both added to them a farder scope, in the termes of the same Artes, & also by hys propre Method, and in peculier termes, procedeth, with helpe of the foresayd Artes, to the performance of complet Experiẽces, which of no particular Art, are hable (Formally) to be challenged . . . And bycause it procedeth by *Experiences*, and searcheth forth the causes of Conclusions, by *Experiences:* and also putteth the Conclusions them selues, in *Experience*, it is named of some *Scientia Experimentalis*. The *Experimentall Science*.[100]

Here the word "experimental" may best be understood as "observational." The concept of a modern controlled experiment does not appear in the "Mathematicall Praeface." Dee pointedly noted that other mathematical disciplines rely to varying degrees on reason, but that "this Art, is no fantasticall Imagination: as some Sophister, might, *Cum suis Insolubilibus*, make a flourish."[101] Yet, although Dee's "Experimentall Science" is frequently noted, this is less an innovation of the author than a reflection of the earlier "experimental" work of Roger Bacon and Nicolaus Cusanus, the two authors to whom Dee referred at this point.

Conclusion

The reader of the "Mathematicall Praeface" may feel something of the urgency with which the piece was written. The diagrammatic folding "Groundplat" indicates the plan followed by the author, but the uneven treatment of the various mathematical arts coupled with frequent digressions convey a certain lack of coherence. The "Mathematical Praeface" would never be listed as one of the masterpieces of Elizabethan prose. Still, this failing does not detract from its interest as a plea for the wider study of all branches of mathematics.

In the "Mathematicall Praeface" Dee explicitly noted the usefulness of mathematics. Whether the subject be medicine, tunnels, gunnery, astrology, or navigation, mathematics is essential for all men. England itself can only benefit by a furtherance of these studies since in this way we may obtain "the Perfection of all Philosophie" which leads to the "most noble State of Common Wealthes."[102] However, beyond this it is evident that for Dee, no less than it had been for Cusanus, mathematics had a more sublime value. An understanding of mathematics leads to a true knowledge of both the Creation and the Creator. Reference is made repeatedly to the fact that Theology requires a deep understanding of all aspects of mathematics.

We may readily grant that Dee's emphasis on mathematics and quantification makes him an interesting and significant figure. But how does he relate to other themes normally considered essential for the rise of modern science? In contrast to some other key figures of the Scientific Revolution Dee saw little need to discard the works of either the ancients or their medieval commentators. Rather, his library catalogs show that he had avidly sought their works. The "Mathematicall Praeface" clearly reflects this appreciation. His references openly acknowledge his debt to ancient, medieval and to contemporary authors.

The "Mathematicall Praeface" is not a call for scientific and educational reform of the sort one would expect to find in Francis Bacon, René Descartes, Paracelsus or J. B. van Helmont. Dee's suggestions for a new approach to the study of nature are less prominent here than they are in the *Propaedeumata Aphoristica* in respect to astrology and in the *Monas Hieroglyphica* in respect to alchemy. In these works a "mathematical" progression of aphorisms, theorems and symbols supposedly lead to a new method for the interpretation of the macrocosm and the microcosm. Thus, although Dee's emphasis on mathematics as necessary for an understanding of nature is surely an important document, it would be dangerous to overemphasize this by extracting from it only the most "modern" statements.

In short, the "Mathematicall Praeface" is best seen in terms of Renaissance scholarship as a whole. To be sure, Dee may be shown as an accomplished mathematician, astronomer and student of navigation. And he did view mathematics as a key to knowledge. Here, like his contemporaries, he turned to Biblical sources to indicate the numerical-quantitative nature of the Creation. In the "Praeface" he placed special emphasis on practical applications of the various arts and when possible he gave technological examples. Dee's discussion of Thaumaturgike or Natural Magic places him in another familiar Renaissance tradition. And although he helped to break new ground in his lengthy quotations from Vitruvius and Archimedes, this should not overshadow the fact that much of his inspiration derived from Lullius, Agrippa and Roger Bacon.

The complexity of assessing Dee's work carries over to the assessment of his influence. During his own lifetime he was renowned as a mathematician of high order. His mastery of Euclid had contributed to his early fame. In addition, however, he was noted as an astronomer, an astrologer, an alchemist, and—to the mob—as a nefarious conjurer. Yet, for all his prominence, Dee's departure for Central Europe in 1583 was to severely affect his fortune in his homeland.

As noted earlier, the "Mathematicall Praeface" was frequently referred to in the late sixteenth century as a major statement of the value of mathematics. Nevertheless, this piece was not to be reprinted until 1651—and then for a final time ten years later. Throughout the first half of the seventeenth century there are occasional references to it. Thomas Browne mentions the "Praeface" because of Dee's discussion of perpetual motion (1646)[103] while John Webster turned to the work as an example of the importance of mathematics in his plea for a Helmontian reform of the universities (1654).[104] Still, among major scientific authors of the seventeenth century we find very few citations. Francis Bacon and Robert Boyle ignore Dee and even the *Polyhistor* (1688, not completed until 1707) of Daniel Georg Morhof—where reference would be expected—is devoid of any.

When we examine Dee's limited list of published works we find that the most frequently reprinted title was not the "Mathematicall Praeface," but the *Monas Hieroglyphica*. Published first in 1564, it was reprinted separately in 1591 and was then included in the second volume of Lazarus Zetzner's monumental *Theatrum Chemicum* in 1602. This collection appeared again in 1613 and in a final edition in 1659. If any of Dee's works may truly be called a classic in the period of the scientific revolution the most likely contender for the honor is this one. Not only did the *Monas Hieroglyphica* go through five editions, it was even to be furnished with its own "key" in the form of a commentary.[105] Thus, as the influence of Dee the mathematician began to wane, the influence of Dee the alchemist was on the ascendant. Concurrently, the publication of sections of Dee's diaries by Meric Casaubon as *A True & Faithful Relation of what passed for many Yeers Between Dr: John Dee . . . and some Spirits* (1659) was to reintroduce Dee to the learned world as a devotee of the black art.

It was this interpretation of Dee as an alchemist and as a magician that was to prevail until the publications of the past four decades by Taylor, Johnson and Yates. But if the earlier view had been marred by the fact that so much of Dee's significant work had been ignored, it is also possible that the revised interpretations have gone too far in their claims for Dee as a scientific prophet. It may be best to recognize Dee as the outstanding sixteenth century English mathematician, but to note him also as a man who hoped to utilize his knowledge as a basis for the reform of the basic sciences of astrology and alchemy. One cannot deny his influence in the late sixteenth century on subjects ranging from navigation and astronomy on the one hand to alchemy and natural magic on the other. However, in the following century his contributions to what we would call the "rational" sciences were gradually to be forgotten while new emphasis was to be placed on his fame as an alchemist and as a *magus*. The history of Dee's reputation is instructive and may well serve as a warning to those who might wish to interpret the Scientific Revolution simply as the growth of positive knowledge accompanied by an almost inevitable decay of the pseudo-sciences.

NOTES

1 E. G. R. Taylor, *The Mathematical Practitioners of Tudor & Stuart England 1485–1714* (Cambridge: For the Institute of Navigation at the University Press, 1968), pp. 170–171.

2 This is developed by Frances A. Yates in her *Theatre of the World* (London: Routledge & Kegan Paul, 1969) and *The Rosicrucian Enlightenment* (London and Boston: Routledge & Kegan Paul, 1972).

3 The basic accounts are to be found in *The Private Diary of Dr. John Dee, and The Catalogue of His Library of Manuscripts, From the Original Manuscripts In the Ashmolean Museum at Oxford, and Trinity College Library, Cambridge*, ed. James Orchard Halliwell [London: Printed for the Camden Society (No. 19) by John Bowyer Nichols and Son, 1842], *Autobiographical Tracts of Dr. John Dee, Warden of the College of Manchester*, ed. James Crossley, *Chetham Miscellanies* (s.l.: The Chetham Society, 1851), vol. 1 [The collection includes *The Compendious Rehearsall* (1592), "Supplication to Queen Mary", "Articles for the Recovery and Preservation of the Ancient Monuments" (1556), "A Necessary Advertizement" (1577), and the *Discourse Apologeticall* (1594–5)]. For the most recent detailed account see Peter J. French, *John Dee. The World of an Elizabethan Magus* (London: Routledge & Kegan Paul, 1972).

4 Dee, *Compendious Rehearsall* in *Autobiographical Tracts*, pp. 5–6.

5 *Ibid.*, p. 5.

6 *Ibid.*

7 *Ibid.*, pp. 6–7.

8 *Ibid.*, pp. 7–8.

9 *Ibid.*, p. 8.

10 French, *op. cit.*, pp. 34–35.

11 *Ibid.*, pp. 8–9.

12 Dee, "A Supplication to Q. Mary . . . for the Recovery and Preservation of Ancient Writers and Monuments" and "Articles Concerning the recovery and preservation of the ancient monuments and old excellent Writers" in *Autobiographical Tracts*, pp. 46–49.

13 M. R. James, *Lists of Manuscripts*

Formerly Owned by Dr. John Dee, Transactions, The Bibliographical Society, London No. 1 (Oxford: O.U.P. for the Bibliographical Society, 1921), p. 10.

14 The printed works are discussed by French, *op. cit.*, pp. 40–61 and Frances Yates makes the contents of Dee's Library a central theme of her *Theatre of the World.*

15 Dee, *Discourse Apologeticall* in *Autobiographical Tracts*, pp. 73–77. A similar list will be found in the *Compendious Rehearsall*, pp. 24–27. However, perhaps the most useful list of Dee's manuscripts and printed works will be found in French, *op. cit.*, pp. 210–217.

16 *Ibid.*, pp. 97–98. See also Francis R. Johnson, *Astronomical Thought in Renaissance England. A Study of the English Scientific Writings from 1500 to 1645* (Baltimore: The Johns Hopkins Press, 1937), pp. 134–135.

17 Dee, *Compendious Rehearsall*, pp. 13–14.

18 Johnson, *op. cit.*, pp. 155–156. Tycho Brahe thought highly of Dee and sent him a copy of his latest work as late as 1590 for his comments. French, *op. cit.*, p. 5.

19 E. G. R. Taylor, *The Haven-Finding Art. A History of Navigation from Odysseus to Captain Cook* (London: Hollis & Carter, 1958), p. 207.

20 French, *op. cit.*, p. 179.

21 *Ibid.*, pp. 93–96.

22 C. H. Josten, "A Translation of John Dee's *Monas Hieroglyphica* (Antwerp, 1564), with an Introduction and Annotations," *Ambix, 12* (1964), 84–221. Josten's work supplants *The Hieroglyphic Monad*, trans. with a commentary by J. W. Hamilton Jones (London: John M. Watkins, 1947). The present discussion is based upon that of the present author to be found in *The Chemical Philosophy. Paracelsian Science and Medicine in the Sixteenth and Seventeenth Centuries* (New York: Science History Publications, in press).

23 Dee, *Monas Hieroglyphica*, trans. Josten, 119, 121.

24 *Ibid.*, 123.

25 *Ibid.*, 125.

26 See Josten's introduction to his translation, *ibid.*, 84–111.

27 *Ibid.*, 175.

28 *Ibid.*, 131.

29 *Ibid.*, 137.

30 Dee, *Compendious Rehearsall* in *Autobiographical Tracts*, p. 19.

31 *Ibid.*, pp. 30–31. Chemical equipment at this time was both expensive and fragile. One is reminded of Robert Boyle's fruitless attempt to send equipment to the family estate at Stalbridge. In a letter to his sister (6 March 1646–7) he wrote that "that great earthen furnace . . . whose conveying hither has taken up so much of my care, and concerning which I made bold very lately to trouble you, since I last did so, has been brought to my hands crumbled into as many pieces, as we into sects; and all the fine experiments, and castles in the air, I had built upon its safe arrival, have felt the fate of their foundation. Well, I see I am not designed to the finding out of the philosophers stone, I have been so unlucky in my first attempts in chemistry. My limbecks, recipients and other glasses have escaped indeed the misfortune of their incendiary, but are now, through the miscarriage of that grand implement of *Vulcan*, as useless to me, as good parts to salvation without the fire of zeal." Thomas Birch, *The Life of the Honourable Robert Boyle* in *The Works of the Honourable Robert Boyle* (6 vols., London: J. and F. Rivington et al., 1772), *1*, pp. v–ccxviii (xxxvi).

32 Dee, *The Private Diary*, pp. 14–15. A still useful early account of Dee and Kelly is to be found in Elias Ashmole (ed.), *Theatrum Chemicum Britannicum* . . . (1652), introduction by Allen G. Debus (New York: Johnson Reprint—Sources of Science No. 39, 1967), pp. 478–484.

33 Dee, *Private Diary*, p. 20.

34 *Ibid.*, p. 21.

35 The extent of this destruction is best described in the *Compendious Rehearsall* in the *Autobiographical Tracts*, pp. 27–31.

36 Dee, *Private Diary*, p. 32.

37 The following discussion closely fol-

lows Allen G. Debus, "Mathematics and Nature in the Chemical Texts of the Renaissance," *Ambix, 15* (1968), 1–28, 211 (3–10).

38 Nicolas Cusanus, *The Idiot in Four Books. The first and second of Wisdome. The third of the Minde. The fourth of statick Experiments, Or Experiments of the Ballance* (London: William Leake, 1650), p. 172. Nicholas Cusanus, *Of Learned Ignorance*, trans. Fr. Germain Heron, O.F.M., Ph.D. (London: Routledge and Kegan Paul, 1954), p. 119.

39 *Ibid.*, p. 118. "With arithmetic the Creator adjusted the World to unity, with geometry he balanced the design to give it stability and controlled movement while with music its parts were so allotted that there should be no more earth in the earth than water in the water, than air in the air or than fire in the fire, so that no element could be wholly transmuted into another: whence it comes that the physical system cannot sink into chaos."

40 *Ibid.*, p. 25. For a comparison of man with nature, Cusanus turned to Plato—"The earth, as Plato says, is like some vast animal whose veins are rivers and whose hairs are the trees; and the animals that feed among the hairs of the earth are as the vermin to be found in the hair of beasts." *Ibid.*, p. 119.

41 *Ibid.*, p. 25.

42 *Ibid.*, p. 26.

43 Debus, "Mathematics and Nature," 4–7.

44 Roger Bacon, *Opus Majus*, trans. Robert Belle Burke (2 vols., Philadelphia, U.P.P., 1928), *1*, p. 261. Such derivations from similar words with radically different meanings are common. Another example may be seen in Sir Christopher Heydon, *A Defence of Iudiciall Astrologie, In Answer to a Treatise lately published by M. Iohn Chamber* (Cambridge: John Legat, 1603), pp. 10–11.

45 Giovanni Pico della Mirandola, *Opera quae extant omnia* (2 vols., Basil: Sebastianum Henricpetri, 1601), *2*, p. 285.

46 On the history of this subject see Karl Sudhoff, *Iatromathematiker vornehmlich im 15. und 16. Jahrhundert* (Breslau: J.U. Kern, 1902).

47 Leonard Digges, *A Prognostication of Right Good Effect, fructfully augmented, contayninge, playne, briefe, pleasant, chosen rules, to iudge the wether for euer, by the Sunne, Moone, Sterres, Cometes, Raynbowe, Thunder, Cloudes, with other Extraordinarie tokens, not omitting the Aspectes of Planets, with a brefe Iudgemene for euer, of Plentie, Lacke, Sicknes, Death, Warres &c . . .* (London: Thomas Gemini, 1555), sig. A4[r]. Digges was a close friend of John Dee.

48 John Baptista Porta, *Natural Magick* (London: Thomas Young and Samuel Speed, 1658), p. 2.

49 *Ibid.*, p. 3.

50 Henricus Cornelius Agrippa, *De Occulta Philosophia. Libri Tres* [(Cologne), 1533], p. 2.

51 *Ibid.*, pp. 99–101.

52 *Ibid.*, p. 160.

53 John Dee, "Mathematicall Praeface" to *The Elements of Geometrie of the most auncient Philosopher Evclide of Megara*, trans. H. Billingsley (London: John Daye, 1570), sig. Aiiii[r].

54 Dee, *Discourse Apologeticall* in the *Autobiographical Tracts,* p. 73.

55 Dee, "Mathematicall Praeface," sig. Aiiii[v] and the folding "Groundplat of my MATHEMATICALL Praeface."

56 Dee, "Mathematicall Praeface," sig. iiii[v].

57 *Ibid.*, sig. *i[r].

58 *Ibid.*, sig. *i[v].

59 *Ibid.*

60 *Ibid.*

61 *Ibid.*, sig. *i[r].

62 *Ibid.*, sig. *ii[r].

63 *Ibid.*, sig. ai[v].

64 *Ibid.*, sigs. *iii[r]–*iiii[v]. This is discussed in some detail in the context of Dee's mathematical philosophy in N. H. Clulee, "John Dee's Mathematics and the Grading of Compound Qualities," *Ambix, 18* (1971), 178–211.

65 Dee, "Mathematicall Praeface," sigs. *iiii[v]–ai[v].

66 *Ibid.*, sig. aiii[r].

Notes.

67 *Ibid.*
68 *Ibid.*, sig. aii^{v}.
69 *Ibid.*, sig. aiii^{r}–aiii^{v}.
70 *Ibid.*, sig. aiii^{v}.
71 *Ibid.*, sig. aiiii^{r}–aiiii^{v}.
72 *Ibid.*, sig. bi^{r}.
73 *Ibid.*, sigs. bi^{v}–bii^{r}.
74 *Ibid.*, sig. bii^{v}.
75 *Ibid.*, sig. biii^{r}.
76 *Ibid.*, sig. biii^{r}–biii^{v}.
77 *Ibid.*, sig. biii^{v}.
78 *Ibid.*
79 *Ibid.*, sig. biiii^{r}.
80 *Ibid.*, sig. biiii^{v}.
81 *Ibid.*, sig. ci^{r}.
82 *Ibid.*, sigs. ciii^{v}–ciiii^{r}.
83 *Ibid.*, sig. ciiii^{r}.
84 *Ibid.*
85 *Ibid.*, sig. di^{r}.
86 *Ibid.*, sig. dii^{r}.
87 *Ibid.*
88 *Ibid.*, dii^{v}.
89 Yates, *Theatre of the World*, pp. 20–41 and *passim.*
90 Dee, "Mathematicall Praeface," sig. diiii^{r}.
91 *Ibid.*
92 *Ibid.*, sig. diiii^{v}.
93 *Ibid.*, sig. Ai^{r}.
94 *Ibid.*
95 *Ibid.*, sig. Ai^{v}.
96 *Ibid.*, sig. Aii^{r}.
97 *Ibid.*
98 *Ibid.*, sig. Aii^{v}.
99 *Ibid.*, sig. Aiii^{r}.
100 *Ibid.*, sigs. Aii^{r}–Aiii^{r}.
101 *Ibid.*, sig. Aiii^{v}.
102 *Ibid.*, sig. Aiiii^{r}.
103 Sir Thomas Browne, *Pseudodoxia Epidemica* in *The Works of Sir Thomas Browne*, ed. Charles Sayle [London: Grant Richards (vols. 1, 2), 1904 and Edinburgh: John Grant (vol. 3), 1907], *2*, p. 253.
104 John Webster, *Academiorum Examen, or the EXAMINATION OF ACADEMIES* in Allen G. Debus, *Science and Education in the Seventeenth Century. The Webster-Ward Debate* (London: Macdonald/New York: American Elsevier, 1970), p. 134.
105 Thomas Tymme, *A LIGHT in Darkness Which illumineth for all the Monas Hieroglyphica of DR. JOHN DEE, discovering Natures closet and revealing the true Christian secrets of Alchimy* (with a note by S. K. Heninger, Jr.) (Oxford: New Bodleian Library, 1963). Tymme's alchemical and Paracelsian interests are discussed in Allen G. Debus, *The English Paracelsians* (London: Oldbourne Press, 1965), pp. 87–89 and *passim.*

ACKNOWLEDGMENTS The author is profoundly grateful to the National Institutes of Health for the necessary financial support required for the completion of this introduction (Research Grant LM-01943-01).

The copy of the "Mathematicall Praeface" reproduced for this edition is from the first English translation of the *Elements* of Euclid (1570) in the Joseph Holly Schaffner Collection of the University of Chicago Libraries. We are deeply indebted to the University of Chicago and specifically to Mr. Robert Rosenthal, Curator of the Department of Special Collections, for permission to photograph this volume.

BIBLIOGRAPHICAL NOTE

The extensive listing of both primary and secondary sources in Peter J. French's *John Dee: The World of an Elizabethan Magus* (London: Routledge & Kegan Paul, 1972) makes unnecessary any detailed discussion of recent works on Dee. Nevertheless, interpretations vary so much that a short resume of twentieth century scholarship—admittedly incomplete—would seem to be of some value at this point.

The biography of Charlotte Fell Smith [*John Dee (1527–1608)* (London: Constable, 1909)], now badly out of date, was responsible for generating a new interest in Dee. Smith's work largely ignored Dee's scientific contribution—and little work was done on the part of historians of science to remedy this situation for many years. Thus, in a chapter on "The Scientific Attitude of the Elizabethans" in his pioneering *Ancients and Moderns: A Study of the Rise of the Scientific Movement in Seventeenth-Century England* (1st ed., 1936; 2nd ed., St. Louis: Washington U.P., 1961) Richard Foster Jones found it possible to ignore Dee. In his work on *The Star-Crossed Renaissance: The Quarrel about Astrology and Its Influence in England* (Durham: Duke U.P., 1941) Don Cameron Allen (p. 105) characterized Dee as "not too far above the lower order of prognosticators." Paul H. Kocher's often useful *Science and Religion in Elizabethan England* (San Marino, Calif.: The

Huntington Library, 1953) unfortunately did little to rectify this interpretation. Here we read (p.158) that Dee had "something of the same blend of mathematics with a sense of the presence of mysterious spiritual vitalities which characterized the later disciples of Plato. He had also, it would appear, very little common sense, much credulity for a glib story, much egoism and showmanship, and not much money."

Modern scholarship on Dee may be said to begin with the research of E. G. R. Taylor. In *Tudor Geography: 1485–1583* (London: Methuen, 1930), *Late Tudor and Early Stuart Geography: 1583–1650* (London: Methuen, 1934), *Mathematical Practitioners of Tudor and Stuart England* (Cambridge: C.U.P. for the Institute of Navigation, 1967)—and in a number of other books and papers—Professor Taylor placed new emphasis on Dee as a major figure in sixteenth century science, geography and navigation. At the same time Francis R. Johnson was reassessing astronomy in England in the sixteenth and the seventeenth centuries. His *Astronomical Thought in Renaissance England: A Study of the English Scientific Writings from 1500 to 1645* (Baltimore: The Johns Hopkins Press, 1937) was to clearly indicate John Dee's knowledge and use of the work of Copernicus.

The study of John Dee's manuscript collection by M. R. James [The Bibliographical Society, London, *Transactions*, no. 1 (1921)] was influential in indicating the breadth of his interests, but it was the work of I. R. F. Calder that was to firmly link Dee with neo-Platonic sources. His *John Dee Studied as an English Neo-Platonist*, an unpublished University of London Dissertation (1952), perhaps may be characterized as a catalyst for the work of a number of students of English Hermeticism in the sixteenth century. Dr. Frances A. Yates has examined the work of Dee both in the *Theatre of the World* (London: Routledge & Kegan Paul, 1970) and in *The Rosicrucian Enlightenment* (London: Routledge & Kegan Paul, 1972). Here she has suggested that Dee not only is responsible for the introduction of the Vitruvian influence in England, but also, that through his sojourn on the continent and his influence on Robert Fludd, Dee played a direct and key role in the rise of modern science. Peter French reflects the deep influence of Dr. Yates in his book on Dee (1972). Also influenced by Yates, but critical of her work on a number of points is N. H. Clulee. His *The Glas of Creation: Renaissance Mathematicism and Natural Philosophy in the Work of John Dee* (1972), a dissertation prepared at the University of Chicago, remains unpublished, but many of his views are to be found in his article, "John Dee's Mathematics and the Grading of Compound Qualities," *Ambix*, *18* (1971), 178–211.

Iohn Dee his Mathematicall Praeface.

Dee's mystical work has always attracted attention. This formed a focal part of Charlotte Fell Smith's biography and it was also the dominant theme of G. N. Hort's *Dr. John Dee, Elizabethan Mystic and Astrologer* (London: Rider, 1922). Little more useful for the historian of science is the recent *John Dee. Scientist, Astrologer and Secret Agent to Elizabeth I* (London: Frederick Muller, 1968). In this work a case is made for Dee as a secret agent in the period 1583–1589. More interesting is Deacon's noting of Robert Hooke's study of John Dee's system of cryptography, but little of value is said about his scientific and mathematical work.

The first English translation of the *Monas Hieroglyphica* was that of J. W. Hamilton Jones [*The Hieroglyphic Monad* (London: Watkins, 1947)], but this has been superseded by the far more thorough edition of C. H. Josten ["A Translation of John Dee's 'Monas Hieroglyphica' (Antwerp, 1564), with an Introduction and Annotations," *Ambix, 12* (1964), 84–221]. Allen G. Debus has placed Dee's *Monas Hieroglyphica* in its chemical-alchemical context in his "Mathematics and Nature in the Chemical Texts of the Renaissance," *Ambix, 15* (1968), 1–28 and also in his *The Chemical Philosophy. Paracelsian Science and Medicine in the Sixteenth and Seventeenth Centuries* (New York: Science History Publications, in press).

Because of the widespread interest aroused by the studies of Frances A. Yates—and to a lesser extent other authors—John Dee will surely become the center of many papers and monographs in the near future. Presently the greatest need is to examine Dee's work—so far as is possible—in its full contemporary context. Any research limited only to Hermeticism or Natural Magic is insufficient. A current problem among historians of science is the widespread unwillingness on the part of many hard core internalists to accept the work of intellectual historians. On the other hand, far too often those historians who are untrained in the sciences seek interpretations that are confined almost exclusively to intellectual, social, religious, or other relationships. Such studies too frequently seem to avoid detailed treatment of the scientific work present in the texts of the authors being examined. If we are to eventually reach a deeper understanding of the Scientific Revolution we will have to integrate the research of both the "internalists" and the "externalists."

The Mathematicall Praeface

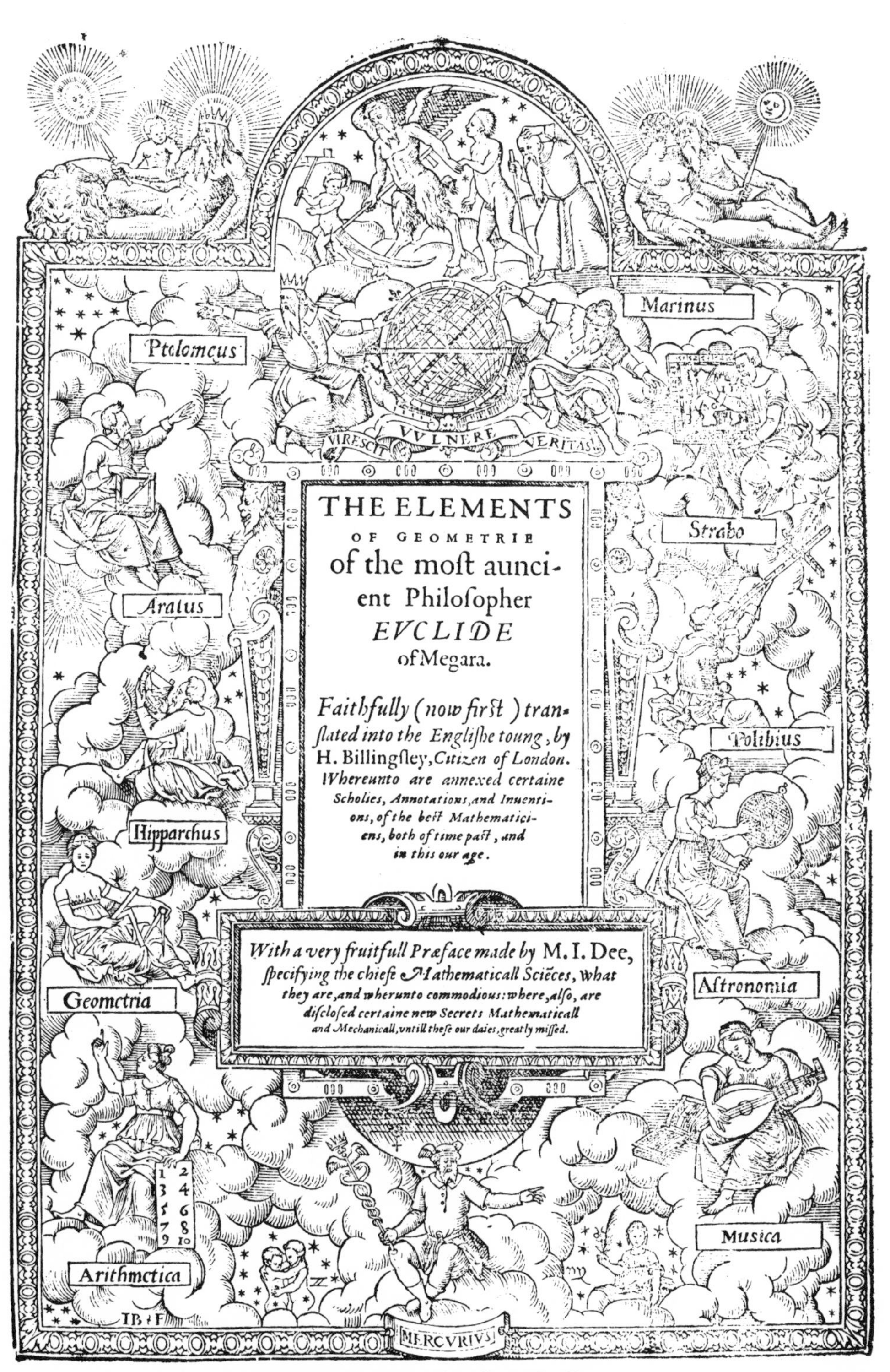

THE ELEMENTS
OF GEOMETRIE
of the most aunci-
ent Philosopher
EVCLIDE
of Megara.

Faithfully (now first) translated into the Englishe toung, by H. Billingsley, *Citizen of London. Whereunto are annexed certaine Scholies, Annotations, and Inuentions, of the best Mathematiciens, both of time past, and in this our age.*

With a very fruitfull Præface made by M. I. Dee, *specifying the chiefe Mathematicall Scieces, what they are, and wherunto commodious: where, also, are disclosed certaine new Secrets Mathematicall and Mechanicall, vntill these our daies, greatly missed.*

Imprinted at London by *Iohn Daye.*

The Tranſlator to the Reader.

Here is (gentle Reader) nothing (the word of God onely ſet apart) which ſo much beautifieth and adorneth the ſoule and minde of mā, as doth the knowledge of good artes and ſciences: as the knowledge of naturall and morall Philoſophie. The one ſetteth before our eyes, the creatures of God, both in the heauens aboue, and in the earth beneath: in which as in a glaſſe, we beholde the exceding maieſtie and wiſedome of God, in adorning and beautifying them as we ſee: in geuing vnto them ſuch wonderfull and manifolde proprieties, and naturall workinges, and that ſo diuerſly and in ſuch varietie: farther in maintaining and conſeruing them continually, whereby to praiſe and adore him, as by S. Paule we are taught. The other teacheth vs rules and preceptes of vertue, how, in common life amongeſt men, we ought to walke vprightly: what dueties pertaine to our ſelues, what pertaine to the gouernment or good order both of an houſholde, and alſo of a citie or common wealth. The reading likewiſe of hiſtories, conduceth not a litle, to the adorning of the ſoule & minde of man, a ſtudie of all men cōmended: by it are ſeene and knowen the artes and doinges of infinite wiſe men gone before vs. In hiſtories are contained infinite examples of heroicall vertues to be of vs followed, and horrible examples of vices to be of vs eſchewed. Many other artes alſo there are which beautifie the minde of man: but of all other none do more garniſhe & beautifie it, then thoſe artes which are called Mathematicall. Vnto the knowledge of which no man can attaine, without the perfecte knowledge and inſtruction of the principles, groundes, and Elementes of Geometrie. But per-

fectly to be instructed in them, requireth diligent studie and reading of olde auncient authors. Amongest which, none for a beginner is to be preferred before the most auncient Philosopher Euclide *of* Megara. *For of all others he hath in a true methode and iuste order, gathered together whatsoeuer any before him had of these Elementes written: inuenting also and adding many thinges of his owne: wherby he hath in due forme accomplished the arte: first geuing definitions, principles, & groundes, wherof he deduceth his Propositions or conclusions, in such wonderfull wise, that that which goeth before, is of necessitie required to the proufe of that which followeth. So that without the diligent studie of* Euclides *Elementes, it is impossible to attaine vnto the perfecte knowledge of Geometrie, and consequently of any of the other Mathematicall sciences. Wherefore considering the want & lacke of such good authors hitherto in our Englishe tounge, lamenting also the negligence, and lacke of zeale to their countrey in those of our nation, to whom God hath geuen both knowledge, & also abilitie to translate into our tounge, and to publishe abroad such good authors, and bookes (the chiefe instrumentes of all learninges): seing moreouer that many good wittes both of gentlemen and of others of all degrees, much desirous and studious of these artes, and seeking for them as much as they can, sparing no paines, and yet frustrate of their intent, by no meanes attaining to that which they seeke: I haue for their sakes, with some charge & great trauaile, faithfully translated into our vulgare tōuge, & set abroad in Print, this booke of* Euclide. *Whereunto I haue added easie and plaine declarations and examples by figures, of the definitions. In which booke also ye shall in due place finde manifolde additions, Scholies, Annotations, and Inuentions: which I haue gathered out of many of the most famous & chiefe Mathematiciēs, both of old time, and in our age: as by diligent reading it in course, ye shall*

well

well perceaue. The fruite and gaine which I require for these my paines and trauaile, shall be nothing els, but onely that thou gentle reader, will gratefully accept the same: and that thou mayest thereby receaue some profite: and moreouer to excite and stirre vp others learned, to do the like, & to take paines in that behalfe. By meanes wherof, our Englishe tounge shall no lesse be enriched with good Authors, then are other straunge tounges: as the Dutch, French, Italian, and Spanishe: in which are red all good authors in a maner, found amongest the Grekes or Latines. Which is the chiefest cause, that amongest thē do florishe so many cunning and skilfull men, in the inuentions of straunge and wonderfull thinges, as in these our daies we see there do. Which fruite and gaine if I attaine vnto, it shall encourage me hereafter, in such like sort to translate, and set abroad some other good authors, both pertaining to religion (as partly I haue already done) and also pertaining to the Mathematicall Artes. Thus gentle reader farewell.

(?¿)

TO THE VNFAINED LOVERS of truthe, and conſtant Studentes of Noble *Sciences,* IOHN DEE *of London, hartily* wiſheth grace from heauen, and moſt proſperous *ſucceſſe in all their honeſt attemptes and* exerciſes.

Iuine *Plato*, the great Maſter of many worthy Philoſophers, and the conſtant auoucher, and pithy perſwader of *Vnum*, *Bonum*, and *Ens*: in his Schole and Academie, ſundry times (beſides his ordinary Scholers) was viſited of a certaine kinde of men, allured by the noble fame of *Plato*, and the great commendation of hys profound and profitable doctrine. But when ſuch Hearers, after long harkening to him, perceaued, that the drift of his diſcourſes iſſued out, to conclude, this *Vnum*, *Bonum*, and *Ens*, to be Spirituall, Infinite, Æternall, Omnipotent, &c. Nothyng beyng alledged or expreſſed, How, worldly goods: how, worldly dignitie: how, health, Strẽgth or luſtines of body: nor yet the meanes, how a merueilous ſenſible and bodyly blyſſe and felicitie hereafter, might be atteyned: Straightway, the fantaſies of thoſe hearers, were dampt: their opinion of *Plato*, was clene chaunged: yea his doctrine was by them deſpiſed: and his ſchole, no more of them viſited. Which thing, his Scholer, *Ariſtotle*, narrowly cōſidering, founde the cauſe therof, to be, For that they had no forwarnyng and information, in generall, whereto his doctrine tended. For, ſo, might they haue had occaſion, either to haue forborne his ſchole hauntyng: (if they, then, had miſliked his Scope and purpoſe) or conſtantly to haue continued therin: to their full ſatiſfaction: if ſuch his finall ſcope & intent, had ben to their deſire. Wherfore, *Ariſtotle*, euer, after that, vſed in brief, to forewarne his owne Scholers and hearers, both of what matter, and alſo to what ende, he tooke in hand to ſpeake, or teach. While I conſider the diuerſe trades of theſe two excellent Philoſophers (and am moſt ſure, both, that *Plato* right well, otherwiſe could teach: and that *Ariſtotle* mought boldely, with his hearers, haue dealt in like ſorte as *Plato* did) I am in no little pang of perplexitie: Bycauſe, that, which I miſlike, is moſt eaſy for me to performe (and to haue *Plato* for my exāple.) And that, which I know to be moſt commendable: and (in this firſt bringyng, into common handling, the *Artes Mathematicall*) to be moſt neceſſary: is full of great difficultie and ſundry daungers. Yet, neither do I think it mete, for ſo ſtraunge matter (as now is ment to be publiſhed) and to ſo ſtraunge an audience, to be bluntly, at firſt, put forth, without a peculiar Preface: Nor (Imitatyng *Ariſtotle*) well can I hope, that accordyng to the amplenes and dignitie of the *State Mathematicall*, I am able, either playnly to preſcribe the materiall boundes: or preciſely to expreſſe the chief purpoſes, and moſt wonderfull applications therof. And though I am ſure, that ſuch as did ſhrinke from *Plato* his ſchole, after they had perceiued his fi-

nall

nall conclusion, would in these thinges haue ben his most diligent hearers) so infinitely mought their desires, in fine and at length, by our *Artes Mathematicall* be satisfied) yet, by this my Præface & forewarnyng, Aswell all such, may (to their great behofe) the soner, hither be allured: as also the *Pythagoricall*, and *Platonicall* perfect scholer, and the constant profound Philosopher, with more ease and spede, may (like the Bee,) gather, hereby, both wax and hony.

Wherfore, seyng I finde great occasion (for the causes alleged, and farder, in re-
„ spect of my *Art Mathematike generall*) to vse a certaine forewarnyng and Præface,
„ whose content shalbe, that mighty, most plesaunt, and frutefull *Mathematicall Tree*,
with his chief armes and second (grifted) braunches: Both, what euery one is, and also, what commodity, in generall, is to be looked for, aswell of griff as stocke: And
„ forasmuch as this enterprise is so great, that, to this our tyme, it neuer was (to my
„ knowledge) by any achieued: And also it is most hard, in these our drery dayes,
„ to such rare and straunge Artes, to wyn due and common credit: Neuertheles, if, for my sincere endeuour to satisfie your honest expectation, you will but lend me your thākefull mynde a while: and, to such matter as, for this time, my penne (with spede) is hable to deliuer, apply your eye or eare attentifely: perchaunce, at once, and for the first salutyng, this Preface you will finde a lesson long enough. And either you will, for a second (by this) be made much the apter: or shortly become, well hable your selues, of the lyons claw, to coniecture his royall symmetrie, and farder propertie. Now then, gentle, my frendes, and countrey men, Turne your eyes, and bend your myndes to that doctrine, which for our present purpose, my simple talent is hable to yeld you.

The intent of this Preface.

All thinges which are, & haue beyng, are found vnder a triple diuersitie generall. For, either, they are demed Supernaturall, Naturall, or, of a third being. Thinges Supernaturall, are immateriall, simple, indiuisible, incorruptible, & vnchangeable. Things Naturall, are materiall, compounded, diuisible, corruptible, and chaungeable. Thinges Supernaturall, are, of the minde onely, comprehended: Things Naturall, of the sense exterior, ar hable to be perceiued. In thinges Naturall, probabilitie and coniecture hath place: But in things Supernaturall, chief demōstration, & most sure Science is to be had. By which properties & comparasons of these two, more easily may be described, the state, condition, nature and property of those thinges, which, we before termed of a third being: which, by a peculier name also, are called *Thynges Mathematicall*. For, these, beyng (in a maner) middle, betwene thinges supernaturall and naturall: are not so absolute and excellent, as thinges supernatural: Nor yet so base and grosse, as things naturall: But are thinges immateriall: and neuerthelesse, by materiall things hable somewhat to be signified. And though their particular Images, by Art, are aggregable and diuisible: yet the generall *Formes*, notwithstandyng, are constant, vnchaungeable, vntrāsformable, and incorruptible. Neither of the sense, can they, at any tyme, be perceiued or iudged. Nor yet, for all that, in the royall mynde of man, first conceiued. But, surmountyng the imperfectiō of coniecture, weenyng and opinion: and commyng short of high intellectuall cōceptiō, are the Mercurial fruite of *Dianœticall* discourse, in perfect imagination subsistyng. A meruaylous newtralitie haue these thinges *Mathematicall*. and also a straunge participatiō betwene thinges supernaturall, immortall, intellectual, simple and indiuisible: and thynges naturall, mortall, sensible, compounded and diuisible. Probabilitie and sensible profe, may well serue in thinges naturall: and is commendable: In Mathematicall reasoninges, a probable Argument, is nothyng regarded: nor yet the testimony of sense, any whit credited: But onely a perfect demonstration, of truthes certaine, necessary, and inuincible: vniuersally and necessaryly concluded:

cluded: is allowed as sufficient for an Argument exactly and purely Mathematical. ››

Of *Mathematicall* thinges, are two principall kindes: namely, *Number*, and *Magnitude*. *Number*, we define, to be, a certayne Mathematicall Sũme, of *Vnits*. And, an *Vnit*, is that thing Mathematicall, Indiuisible, by participation of some likenes of whose property, any thing, which is in deede, or is counted One, may resonably be called One. We account an *Vnit*, a thing *Mathematicall*, though it be no *Number*, and also indiuisible: because, of it, materially, Number doth consist: which, principally, is a thing *Mathematicall*. *Magnitude* is a thing *Mathematicall*, by participation of some likenes of whose nature, any thing is iudged long, broade, or thicke. A thicke *Magnitude* we call a *Solide*, or a *Body*. What *Magnitude* so euer, is Solide or Thicke, is also broade, & long. A broade magnitude, we call a *Superficies* or a Plaine. Euery playne magnitude, hath also length. A long magnitude, we terme a *Line*. A *Line* is neither thicke nor broade, but onely long: Euery certayne Line, hath two endes: The endes of a line, are *Pointes* called. A *Point*, is a thing *Mathematicall*, indiuisible, which may haue a certayne determined situation. If a Poynt moue from a determined situation, the way wherein it moued, is also a *Line*: mathematically produced. whereupon, of the auncient Mathematiciens, a *Line* is called the race or course of a *Point*. A Poynt we define, by the name of a thing Mathematicall: though it be no Magnitude, and indiuisible: because it is the propre ende, and bound of a *Line*: which is a true *Magnitude*. And *Magnitude* we may define to be that thing *Mathematicall*, which is diuisible for euer, in partes diuisible, long, broade or thicke. Therefore though a Poynt be no *Magnitude*, yet *Terminatiuely* we recken it a thing *Mathematicall* (as I sayd) by reason it is properly the end, and bound of a line.

Number.

Note the worde, Vnit, to expresse the Greke Monas, & not Vnitie: as we haue all, commonly, till now vsed.

Magnitude. ››

A point. ››

A Line.

Magnitude.

Neither *Number*, nor *Magnitude*, haue any Materialitie. First, we will consider of *Number*, and of the Science *Mathematicall*, to it appropriate, called *Arithmetike*: and afterward of *Magnitude*, and his Science, called *Geometrie*. But that name contenteth me not: whereof a word or two hereafter shall be sayd. How Immateriall and free from all matter, *Number* is, who doth not perceaue? yea, who doth not wonderfully wõder at it? For, neither pure *Element*, nor *Aristoteles*, *Quinta Essentia*, is hable to serue for Number, as his propre matter. Nor yet the puritie and simplenes of Substance Spirituall or Angelicall, will be found propre enough thereto. And therefore the great & godly Philosopher *Anitius Boetius*, sayd: ***Omnia quæcunq; a primæua rerum natura constructa sunt, Numerorum videntur ratione formata. Hoc enim fuit principale in animo Conditoris Exemplar.*** That is: *All thinges (which from the very first originall being of thinges, haue bene framed and made) do appeare to be Formed by the reason of Numbers. For this was the principall example or patterne in the minde of the Creator.* O comfortable allurement, O rauishing perswasion, to deale with a Science, whose Subiect, is so Auncient, so pure, so excellent, so surmounting all creatures, so vsed of the Almighty and incomprehensible wisdome of the Creator, in the distinct creation of all creatures: in all their distinct partes, properties, natures, and vertues, by order, and most absolute number, brought, from *Nothing*, to the *Formalitie* of their being and state. By *Numbers* propertie therefore, of vs, by all possible meanes, (to the perfection of the Science) learned, we may both winde and draw our selues into the inward and deepe search and vew, of all creatures distinct vertues, natures, properties, and *Formes*: And also, farder, arise, clime, ascend, and mount vp (with Speculatiue winges) in spirit, to behold in the Glas of Creation, the *Forme* of *Formes*, the *Exemplar Number* of all thinges *Numerable*: both visible and inuisible: mortall and immortall, Corporall and Spirituall. Part of this profound and diuine Science, had *Ioachim* the Prophesier atteyned vnto: by *Numbers Formall, Naturall,* and *Rationall,* forseyng, concludyng, and forshewyng great particular euents, long before their comming. His bookes yet remainyng, hereof, are good profe: And the noble Earle of *Mirandula,* (besides that,) a sufficient witnesse: that *Ioachim, in his prophesies, proceded by no other way, then by Numbers Formall.* And this Earle hym selfe, in Rome, *set vp 900. Conclusions, in all kinde of Sciences, openly to be disputed of: and among the rest, in his Conclusions *Mathematicall,* (in the eleuenth Conclusion) hath in Latin, this English sentence. *By Numbers, a way is had, to the searchyng out, and vnderstandyng of euery thyng, hable to be knowen. For the verifying of which Conclusion, I promise to aunswere to the 74. Quæstions, vnder written, by the way of Numbers.* Which Cõclusions, I omit here to rehearse: aswell auoidyng superfluous prolixitie: as, bycause *Ioannes Picus, workes,* are commonly had. But, in any case, I would wish that those Conclusions were red diligently, and perceiued of such, as are earnest Obseruers and Considerers of the constant law of nũbers: which is planted in thyngs Naturall and Supernaturall: and is prescribed to all Creatures, inuiolably to be kept. For, so, besides many other thinges, in those Conclusions to be marked, it would apeare, how sincerely, & within my boundes, I disclose the wonderfull mysteries, by numbers, to be atteyned vnto.

Ano. 1488.

Of my former wordes, easy it is to be gathered, that *Number* hath a treble state: One, in the Creator: an other in euery Creature (in respect of his complete constitution:) aud the third, in Spirituall and Angelicall Myndes, and in the Soule of mã. In the first and third state, *Number,* is termed *Number Numbryng.* But in all Creatures, otherwise, *Number,* is termed *Nũber Numbred.* And in our Soule, Nũber beareth such a swaye, and hath such an affinitie therwith: that some of the old *Philosophers* taught, *Mans Soule, to be a Number mouyng it selfe.* And in dede, in vs, though it be a very Accident: yet such an Accident it is, that before all Creatures it had perfect beyng, in the Creator, Sempiternally. *Number Numbryng* therfore, is the discretion discerning, and distincting of thinges. But in God the Creator, This discretion, in the beginnyng, produced orderly and distinctly all thinges. For his *Numbryng,* then, was his Creatyng of all thinges. And his Continuall *Numbryng,* of all thinges, is the Conseruation of them in being: And, where and when he will lacke an *Vnit:* there and then, that particular thyng shalbe *Discreated.* Here I stay. But our Seuerallyng, distinctyng, and *Numbryng,* createth nothyng: but of Multitude considered, maketh certaine and distinct determination. And albeit these thynges be waighty and truthes of great importance, yet (by the infinite goodnes of the Almighty *Ternarie,*) Artificiall Methods and easy wayes are made, by which the zelous Philosopher, may wyn nere this Riuerish *Ida,* this Mountayne of Contemplation: and more then Contemplation. And also, though *Number,* be a thyng so Immateriall, so diuine, and æternall: yet by degrees, by litle and litle, stretchyng forth, and applying some likenes of it, as first, to thinges Spirituall: and then, bryngyng it lower, to thynges sensibly perceiued: as of a momentanye sounde iterated: then to the least thynges that may be seen, numerable: And at length, (most grossely,) to a multitude of any corporall thynges seen, or felt: and so, of these grosse and sensible thynges, we are trayned to learne a certaine Image or likenes of numbers: and to vse Arte in them to our pleasure and proffit. So grosse is our conuersation, and dull is our apprehension: while mortall Sense, in vs, ruleth the common wealth of our litle world. Hereby we say, Three Lyons, are three: or a *Ternarie.* Three Egles, are three, or a *Ternarie.* Which* *Ternaries,* are eche, the *Vnion, knot,* and *Vniformitie,* of three discrete and distinct *Vnits.* That is, we may in eche *Ternarie,* thrise, seuerally pointe, and shew a part, *One, One,* and *One.* Where, in Numbryng, we say One, two, Three,

Three . But how farre,theſe viſible Ones , do differre from our Indiuiſible Vnits (in pure *Arithmetike*,principally conſidered)no man is ignorant . Yet from theſe groſſe and materiall thynges,may we be led vpward,by degrees,ſo,informyng our rude Imagination,toward the cõceiuyng of *Numbers*,abſolutely(:Not ſuppoſing, nor admixtyng any thyng created,Corporall or Spirituall,to ſupport,conteyne,or repreſent thoſe *Numbers* imagined:) that at length, we may be hable, to finde the number of our owne name , gloriouſly exemplified and regiſtred in the booke of the *Trinitie* moſt bleſſed and æternall.

But farder vnderſtand,that vulgar Practiſers,haue Numbers , otherwiſe, in ſundry Conſiderations:and extend their name farder,then to Numbers , whoſe leaſt part is an *Vnit*.For the common Logiſt,Reckenmaſter, or Arithmeticien, in hys vſing of Numbers:of an Vnit,imagineth leſſe partes:and calleth them *Fractions*. As of an *Vnit* , he maketh an halfe,and thus noteth it,$\frac{1}{2}$.and ſo of other,(infinitely diuerſe)partes of an *Vnit*.Yea and farder,hath, *Fractions of Fractions. &c* . And,foraſmuch,as, *Addition , Subſtraction , Multiplication, Diuiſion* and *Extraction of Rotes*,are the chief,and ſufficient partes of *Arithmetike* : which is , the *Science that demonſtrateth the properties,of Numbers,and all operatiõs , in numbers to be performed:* How often, therfore,theſe fiue ſundry ſortes of Operations, do , for the moſt part,of their execution,differre from the fiue operations of like generall property and name,in our Whole numbers practiſable,So often , (for a more diſtinct doctrine) we,vulgarly account and name it,an other kynde of *Arithmetike*. And by this reaſon:the Conſideration,doctrine,and working,in whole numbers onely: where, of an *Vnit*,is no leſſe part to be allowed:is named(as it were)an *Arithmetike* by it ſelfe . And ſo of the *Arithmetike of Fractions*.In lyke ſorte,the neceſſary,wonderfull and Secret doctrine of Proportion , and proportionalytie hath purchaſed vnto it ſelfe a peculier maner of handlyng and workyng:and ſo may ſeme an other forme of *Arithmetike*. Moreouer,the *Aſtronomers*,for ſpede and more commodious calculation,haue deuiſed a peculier maner of orderyng nũbers,about theyr circular motions, by Sexagenes,and Sexageſmes.By Signes,Degrees and Minutes &c. which commonly is called the *Arithmetike* of *Aſtronomical* or *Phiſicall Fractions*.That, haue I briefly noted,by the name of *Arithmetike Circular*. Bycauſe it is alſo vſed in circles,not *Aſtronomicall.&c*.Practiſe hath led *Numbers* farder , and hath framed them,to take vpon them , the ſhew of *Magnitudes* propertie: Which is *Incommenſurabilitie* and *Irrationalitie*. (For in pure *Arithmetike*,an *Vnit*,is the common Meaſure of all Numbers.) And,here,Nũbers are become,as Lynes,Playnes and Solides: ſome tymes *Rationall*,ſome tymes *Irrationall*. And haue propre and peculier characters,(as √ȝ.√ce. and ſo of other. Which is to ſignifie *Rote Square , Rote Cubik:and ſo forth*:)& propre and peculier faſhions in the fiue principall partes: Wherfore the practiſer,eſtemeth this,a diuerſe *Arithmetike* from the other . Practiſe bryngeth in,here,diuerſe compoundyng of Numbers: as ſome tyme,two,three,foure(or more) *Radicall* nũbers, diuerſly knit,by ſignes, of More & Leſſe:as thus √ȝ. 12 + √ce 15.Or thus √ȝȝ. 19 + √ce 12—√ȝ. 2. &c.And ſome tyme with whole numbers, or fractions of whole Number,amõg them:as 20 + √ȝ.24.√ce16 + 33—√ȝ.10.√ȝȝ.44 + 12 ¼ + √ce9. And ſo, infinitely , may hap the varietie. After this : Both the one and the other hath fractions incident:and ſo is this *Arithmetike* greately enlarged,by diuerſe exhibityng and vſe of Compoſitions and mixtynges . Conſider how, I(beyng deſirous to deliuer the ſtudent from error and Cauillation)do giue to this *Practiſe*,the name of the *Arithmetike of Radicall numbers*: Not,of *Irrationall* or *Surd Numbers*: which other while, are Rationall : though they haue the Signe of a Rote before

Arithmetike.

Note.

1\. 2. 3. 4.

them,

them, which, *Arithmetike* of whole Numbers most vsuall, would say they had no such Roote: and so account them *Surd Numbers*: which, generally spokē, is vntrue: as *Euclides* tenth booke may teach you. Therfore to call them, generally, *Radicall Numbers*, (by reason of the signe √. prefixed,) is a sure way: and a sufficient generall distinction from all other ordryng and vsing of Numbers: And yet (beside all this) Consider: the infinite desire of knowledge, and incredible power of mans Search and Capacitye: how, they, ioyntly haue waded farder (by mixtyng of speculation and practise) and haue found out, and atteyned to the very chief perfection (almost) of *Numbers* Practicall vse. Which thing, is well to be perceiued in that great Arithmeticall Arte of *Æquation*: commonly called the *Rule of Coss*. or *Algebra*. The Latines termed it, *Regulam Rei & Census*, that is, the *Rule of the thyng and his value*. With an apt name: comprehendyng the first and last pointes of the worke. And the vulgar names, both in Italian, Frenche and Spanish, depend (in namyng it,) vpon the signification of the Latin word, *Res*: *A thing*: vnleast they vse the name of *Algebra*. And therin (commonly) is a dubble error. The one, of them, which thinke it to be of *Geber* his inuentyng: the other of such as call it *Algebra*. For, first, though *Geber* for his great skill in Numbers, Geometry, Astronomy, and other maruailous Artes, mought haue semed hable to haue first deuised the sayd Rule: and also the name carryeth with it a very nere likenes of *Geber* his name: yet true it is, that a *Greke* Philosopher and Mathematicien, named *Diophantus*, before *Geber* his tyme, wrote 13. bookes therof (of which, six are yet extant: and I had them to * vse, of the famous Mathematicien, and my great frende, *Petrus Montaureus*:) And secondly, the very name, is *Algiebar*, and not *Algebra*: as by the Arabien *Auicen*, may be proued: who hath these precise wordes in Latine, by *Andreas Alpagus* (most perfect in the Arabik tung) so translated. *Scientia faciendi Algiebar & Almachabel*. i. *Scientia inueniendi numerum ignotum, per additionem Numeri, & diuisionem & æquationem*. Which is to say: *The Science of workyng Algiebar and Almachabel*, that is, the *Science of findyng an vnknowen number, by Addyng of a Number, & Diuision & æquation*. Here haue you the name: and also the principall partes of the Rule, touched. To name it, *The rule, or Art of Æquation*, doth signifie the middle part and the State of the Rule. This Rule, hath his peculier Characters: and the principal partes of *Arithmetike*, to it appertayning, do differre from the other *Arithmeticall operations*. This *Arithmetike*, *hath Nūbers* Simple, Cōpound, Mixt: and Fractions, accordingly. This Rule, and *Arithmetike* of *Algiebar*, is so profound, so generall and so (in maner) conteyneth the whole power of Numbers Application practicall: that mans witt, can deale with nothyng, more proffitable about numbers: nor match, with a thyng, more mete for the diuine force of the Soule, (in humane Studies, affaires, or exercises) to be tryed in. Perchaunce you looked for, (long ere now,) to haue had some particular profe, or euident testimony of the vse, proffit and Commodity of Arithmetike vulgar, in the Common lyfe and trade of men. Therto, then, I will now frame my selfe: But herein great care I haue, least length of sundry profes, might make you deme, that either I did misdoute your zelous mynde to vertues schole: or els mistrust your hable witts, by some, to gesse much more. A profe then, foure, fiue, or six, such, will I bryng, as any reasonable man, therwith may be persuaded, to loue & honor, yea learne and exercise the excellent Science of *Arithmetike*.

*Anno. 1550.

5.

And first: who, nerer at hand, can be a better witnesse of the frute receiued by *Arithmetike*, then all kynde of Marchants? Though not all, alike, either nede it, or vse it. How could they forbeare the vse and helpe of the Rule, called the Golden Rule?

Rules: Simple and Compounde: both forward and backward? How might they misse *Arithmeticall* helpe in the Rules of Felowshyp: either without tyme, or with tyme? and betwene the Marchant & his Factor? The Rules of Bartering in wares onely: or part in wares, and part in money, would they gladly want? Our Marchant venturers, and Trauaylers ouer Sea, how could they order their doynges iustly and without losse, vnleast certaine and generall Rules for Exchaũge of money, and Rechaunge, were, for their vse, deuised? The Rule of Alligation, in how sundry cases, doth it conclude for them, such precise verities, as neither by naturall witt, nor other experience, they, were hable, els, to know? And (with the Marchant then to make an end) how ample & wonderfull is the Rule of False positions? especially as it is now, by two excellent Mathematiciens (of my familier acquayntance in their life time) enlarged? I meane *Gemma Frisius*, and *Simon Iacob*. Who can either in brief conclude, the generall and Capitall Rules? or who can Imagine the Myriades of sundry Cases, and particular examples, in Act and earnest, continually wrought, tried and concluded by the forenamed Rules, onely? How sundry other *Arithmeticall practises*, are commonly in Marchantes handes, and knowledge: They them selues, can, at large, testifie.

The Mintmaster, and Goldsmith, in their Mixture of Metals, either of diuerse kindes, or diuerse values: how are they, or may they, exactly be directed, and meruailously pleasured, if *Arithmetike* be their guide? And the honorable Phisiciãs, will gladly confesse them selues, much beholding to the Science of *Arithmetike*, and that sundry wayes: But chiefly in their Art of Graduation, and compounde Medicines. And though *Galenus, Auerrois, Arnoldus, Lullus*, and other haue published their positions, aswell in the quantities of the Degrees aboue Temperament, as in the Rules, concluding the new *Forme* resulting: yet a more precise, commodious, and easy *Method*, is extant: by a Countreyman of ours (aboue 200. yeares ago) inuented. And forasmuch as I am vncertaine, who hath the same: or when that litle Latin treatise, (as the Author writ it,) shall come to be Printed: (Both to declare the desire I haue to pleasure my Countrey, wherin I may: and also, for very good profe of Numbers vse, in this most subtile and frutefull, Philosophicall Conclusion,) I entend in the meane while, most briefly, and with my farder helpe, to communicate the pith therof vnto you.

R. B.

First describe a circle: whose diameter let be an inch. Diuide the Circumference into foure equall partes. Frõ the Center, by those 4. sections, extend 4. right lines: eche of 4. inches and a halfe long: or of as many as you liste, aboue 4. without the circumference of the circle: So that they shall be of 4. inches long (at the least) without the Circle. Make good euident markes, at euery inches end. If you list, you may subdiuide the inches againe into 10. or 12. smaller partes, equall. At the endes of the lines, write the names of the 4. principall elementall Qualities. *Hote* and *Colde*, one against the other. And likewise *Moyst* and *Dry*, one against the other. And in the Circle write *Temperate*. Which *Temperature* hath a good Latitude: as appeareth by the Complexion of man. And therefore we haue allowed vnto it, the foresayd Circle: and not a point Mathematicall or Physicall.

Now, when you haue two thinges Miscible, whose degrees are * truely knowen: Of necessitie, either they are of one Quantitie and waight, or of diuerse. If they be of one Quantitie and waight: whether their formes, be Contrary Qualities, or of one kinde (but of diuerse intentions and degrees) or a *Temperate*, and a Contrary, *The forme resulting of their Mixture, is in the Middle betwene the degrees of*

**Take some part of Lullus counsayle in his booke de Q. Essentia.*

the formes mixt. As for example,let *A*,be *Moist* in the first degree : and *B*, *Dry* in the third degree. Adde 1. and 3. that maketh 4 : the halfe or middle of 4. is 2. This 2. is the middle, equally distant from *A* and *B* (for the* *Temperament* is counted none. And for it, you must put a Ciphre, if at any time, it be in mixture).

*Note.

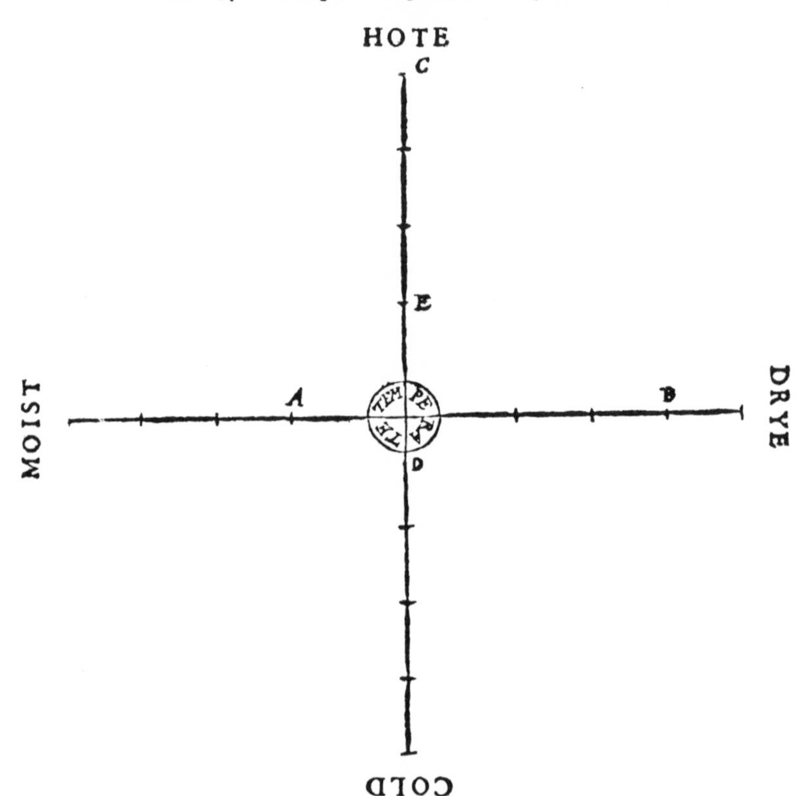

Counting then from *B*, 2. degrees, toward *A*: you finde it to be *Dry* in the first degree : So is the *Forme resulting* of the Mixture of *A*, and *B*, in our example. I will geue you an other example. Suppose, you haue two thinges, as *C*, and *D* : and of *C*, the Heate to be in the 4. degree : and of *D*, the Colde, to be remisse, euen vnto the *Temperament*. Now, for *C*, you take 4: and for *D*, you take a Ciphre : which, added vnto 4, yeldeth onely 4. The middle, or halfe, whereof, is 2. Wherefore the *Forme resulting* of *C*, and *D*, is Hote in the second degree: for, 2. degrees, accounted from *C*, toward *D*, ende iuste in the 2. degree of heate. Of the third maner, I will geue also an example: which let be this : I haue a liquid Medicine whose Qualitie of heate is in the 4. degree exalted : as was *C*, in the example foregoing: and an other liquid Medicine I haue : whose Qualitie, is heate, in the first degree. Of eche of these, I mixt a like quantitie : Subtract here, the lesse frõ the more : and the residue diuide into two equall partes : whereof, the one part, either added to the lesse, or subtracted from the higher degree, doth produce the degree of the Forme

Note.

Forme resulting,by this mixture of *C*,and *E*. As,if from 4. ye abate 1.there resteth 3.the halfe of 3. is 1$\frac{1}{2}$: Adde to 1.this 1$\frac{1}{2}$: you haue 2$\frac{1}{2}$. Or subtract from 4. this 1$\frac{1}{2}$: you haue likewise 2$\frac{1}{2}$ remayning. Which declareth, the *Forme resulting*, to be *Heate*, in the middle of the third degree.

The Second Rule.

But if the Quantities of two thinges Commixt, be diuerse, and the Intensions (of their Formes Miscible) be in diuerse degrees, and heigthes. (Whether those Formes be of one kinde, or of Contrary kindes, or of a Temperate and a Contrary, *What proportion is of the lesse quantitie to the greater, the same shall be of the difference, which is betwene the degree of the Forme resulting, and the degree of the greater quantitie of the thing miscible, to the difference, which is betwene the same degree of the Forme resulting, and the degree of the lesse quantitie*. As for example. Let two pound of Liquor be geuen, hote in the 4.degree: & one pound of Liquor be geuen, hote in the third degree. I would gladly know the Forme resulting, in the Mixture of these two Liquors. Set downe your nũbers in order, thus.

℔. 2.	*Hote.* 4.
℔. 1.	*Hote.* 3.

Now by the rule of Algiebar, haue I deuised a very easie, briefe, and generall maner of working in this case. Let vs first, suppose that *Middle Forme resulting*, to be 1ꝶ : as that Rule teacheth. And because (by our Rule, here geuen) as the waight of 1.is to 2 : So is the difference betwene 4.(the degree of the greater quantitie) and 1ꝶ : to the difference betwene 1ꝶ and 3: (the degree of the thing, in lesse quãtitie. And with all, 1ꝶ, being alwayes in a certaine middell, betwene the two heigthes or degrees). For the first difference, I set 4—1ꝶ: and for the second, I set 1ꝶ—3. And, now againe, I say, as 1.is to 2.so is 4—1ꝶ to 1ꝶ—3. Wherfore, of these foure proportionall numbers, the first and the fourth Multiplied, one by the other, do make as much, as the second and the third Multiplied the one by the other. Let these Multiplications be made accordingly. And of the first and the fourth, we haue 1ꝶ—3.and of the second & the third, 8—2ꝶ.Wherfore, our Æquation is betwene 1ꝶ—3: and 8—2ꝶ.Which may be reduced, according to the Arte of Algiebar: as, here, adding 3.to eche part, geueth the Æquation, thus, 1ꝶ=11—2ꝶ. And yet againe, contracting, or Reducing it: Adde to eche part, 2ꝶ: Then haue you 3ꝶ æquall to 11 : thus represented 3ꝶ=11. Wherefore, diuiding 11.by 3: the Quotient is 3$\frac{2}{3}$: the *Valew* of our 1ꝶ, *Coß*, or *Thing*, first supposed. And that is the heigth, or Intension of the *Forme resulting* : which is, *Heate*, in two thirdes of the fourth degree : And here I set the shew of the worke in conclusion, thus. The proufe hereof is easie by subtracting 3.from 3$\frac{2}{3}$, resteth $\frac{2}{3}$. Subtracte the same heigth of the Forme resulting, (which is 3$\frac{2}{3}$) frõ 4: then resteth $\frac{1}{3}$: You see, that $\frac{2}{3}$ is double to $\frac{1}{3}$:

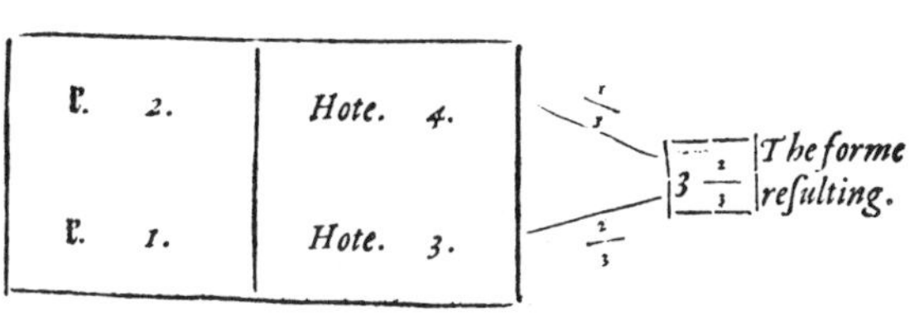

as 2. ℔. is double to 1. ℔. So should it be : by the rule here geuen. Note. As you added to eche part of the Æquation, 3 : so if ye first added to eche part 2ꝶ, it would stand, 3ꝶ—3=8. And now adding to eche part 3 : you haue (as afore) 3ꝶ=11.

And though I, here, speake onely of two thyngs Miscible: and most commonly, mo then three, foure, fiue or six, (&c.) are to be Mixed: (and in one Compound to be reduced:& the Forme resultyng of the same, to serue the turne)yet these Rules are sufficient: duely repeated and iterated. In procedyng first, with any two: and then, with the Forme Resulting, and an other: & so forth: For, the last worke, concludeth the Forme resultyng of them all: I nede nothing to speake, of the Mixture (here supposed) what it is. Common Philosophie hath defined it, saying, *Mixtio est miscibilium, alteratorum, per minima coniunctorum, Vnio.* Euery word in the definition, is of great importance. I nede not also spend any time, to shew, how, the other manner of distributing of degrees, doth agree to these Rules. Neither nede I of the farder vse belonging to the Crosse of Graduation (before described) in this place declare, vnto such as are capable of that, which I haue all ready sayd. Neither yet with examples specifie the Manifold varieties, by the foresayd two generall Rules, to be ordered. The witty and Studious, here, haue sufficient: And they which are not hable to atteine to this, without liuely teaching, and more in particular: would haue larger discoursing, then is mete in this place to be dealt withall: And other (perchaunce) with a proude snuffe will disdaine this litle: and would be vnthankefull for much more. I, therfore conclude: and wish such as haue modest and earnest Philosophicall mindes, to laude God highly for this: and to Meruayle, that the profoundest and subtilest point, concerning *Mixture of Formes and Qualities Naturall*, is so Matcht and maryed with the most simple, easie, and short way of the noble Rule of *Algiebar*. Who can remaine, therfore vnpersuaded, to loue, alow, and honor the excellent Science of *Arithmetike*? For, here, you may perceiue that the litle finger of *Arithmetike*, is of more might and contriuing, then a hunderd thousand mens wittes, of the middle sorte, are hable to perfourme, or truely to conclude, with out helpe thereof.

Note.

Now will we farder, by the wise and valiant Capitaine, be certified, what helpe he hath, by the Rules of *Arithmetike*: in one of the Artes to him appertaining: And „ of the Grekes named Ταχτικὴ. That is, the Skill of Ordring Souldiers in Battell ray „ after the best maner to all purposes. This Art so much dependeth vppon Numbers vse, and the Mathematicals, that *Ælianus* (the best writer therof,) in his worke, to the *Emperour Hadrianus*, by his perfection, in the Mathematicals, (beyng greater, then other before him had,) thinketh his booke to passe all other the excellent workes, written of that Art, vnto his dayes. For, of it, had written *Æneas*: *Cyneas* of *Thessaly*: *Pyrrhus Epirota*: and *Alexander* his sonne: *Clearchus*: *Pausanias*: *Euangelus*: *Polybius*, familier frende to *Scipio*: *Eupolemus*: *Iphicrates*, *Possidonius*: and very many other worthy Capitaines, Philosophers and Princes of Immortall fame and memory: Whose fayrest floure of their garland (in this feat) was *Arithmetike*: and a litle perceiuerance, in *Geometricall* Figures. But in many other cases doth *Arithmetike* stand the Capitaine in great stede. As in proportionyng of vittayles, for the Army, either remaining at a stay: or suddenly to be encreased with a certaine number of Souldiers: and for a certain tyme. Or by good Art to diminish his company, to make the victuals, longer to serue the remanent, & for a certaine determined tyme: if nede so require. And so in sundry his other accountes, Reckeninges, Measurynges, and proportionynges, the wise, expert, and Circumspect Capitaine will affirme the Science of *Arithmetike*, to be one of his chief Counsaylors, directers and aiders. Which thing (by good meanes) was euident to the Noble, ☞ the Couragious, the loyall, and Curteous *Iohn*, late Earle of Warwicke. Who was a yong Gentleman, throughly knowne to very few. Albeit his lusty valiantnes, force, and Skill in Chiualrous feates and exercises: his humblenes, and frendelynes to all men, were thinges, openly, of the world perceiued. But what rotes (otherwise,) vertue had fastened in his brest, what Rules of godly and honorable

Ταχτικὴ.

life

life he had framed to him selfe: what vices, (in some then liuing) notable, he tooke great care to eschew: what manly vertues, in other noble men, (florishing before his eyes,) he Sythingly aspired after: what prowesses he purposed and ment to achieue: with what feats and Artes, he began to furnish and fraught him selfe, for the better seruice of his Kyng and Countrey, both in peace & warre. These (I say) his Heroicall Meditations, forecastinges and determinations, no twayne, (I thinke) beside my selfe, can so perfectly, and truely report. And therfore, in Conscience, I count it my part, for the honor, preferment, & procuring of vertue (thus, briefly) to haue put his Name, in the Register of *Fame Immortall.*

To our purpose. This *Iohn*, by one of his actes (besides many other: both in England and Fraunce, by me, in him noted.) did disclose his harty loue to vertuous Sciences: and his noble intent, to excell in Martiall prowesse: When he, with humble request, and instant Solliciting: got the best Rules (either in time past by Greke or Romaine, or in our time vsed: and new Stratagemes therin deuised) for ordring of all Companies, summes and Numbers of mẽ, (Many, or few) with one kinde of weapon, or mo, appointed: with Artillery, or without: on horsebacke, or on fote: to giue, or take onset: to seem many, being few: to seem few, being many. To marche in battaile or Iornay: with many such feates, to Foughten field, Skarmoush, or Ambushe appartaining: And of all these, liuely designementes (most curiously) to be in velame parchement described: with Notes & peculier markes, as the Arte requireth: and all these Rules, and descriptions Arithmeticall, inclosed in a riche Case of Gold, he vsed to weare about his necke: as his Iuell most precious, and Counsaylour most trusty. Thus, *Arithmetike*, of him, was shryned in gold: Of *Numbers* frute, he had good hope. Now, Numbers therfore innumerable, in *Numbers* prayse, his shryne shall finde.

This noble Earle, dyed Anno. 1554. skarse of 24. yeares of age: hauing no issue by his wife: Daughter to the Duke of Somerset.

What nede I, (for farder profe to you) of the Scholemasters of Iustice, to require testimony: how nedefull, how frutefull, how skillfull a thing *Arithmetike* is? I meane, the Lawyers of all sortes. Vndoubtedly, the Ciuilians, can meruaylously declare: how, neither the Auncient Romaine lawes, without good knowledge of *Numbers art*, can be perceiued: Nor (Iustice in infinite Cases) without due proportion, (narrowly considered,) is hable to be executed. How Iustly, & with great knowledge of Arte, did *Papinianus* institute a law of partition, and allowance, betwene man and wife after a diuorce? But how *Accursius, Baldus, Bartolus, Iason, Alexander*, and finally *Alciatus*, (being otherwise, notably well learned) do iumble, gesse, and erre, from the æquity, art and Intent of the lawmaker: *Arithmetike* can detect, and conuince: and clerely, make the truth to shine. Good *Bartolus*, tyred in the examining & proportioning of the matter: and with *Accursius* Glosse, much cumbred: burst out, and sayd: *Nulla est in toto libro, hac glossa difficilior: Cuius computationem nec Scholastici nec Doctores intelligunt. &c.* That is: *In the whole booke, there is no Glosse harder then this: Whose accoumpt or reckenyng, neither the Scholers, nor the Doctours vnderstand. &c.* What can they say of *Iulianus* law, *Si ita Scriptum. &c.* Of the Testators will iustly performing, betwene the wife, Sonne and daughter? How can they perceiue the æquitie of *Aphricanus*, *Arithmeticall* Reckening, where he treateth of *Lex Falcidia*? How can they deliuer him, from his Reprouers: and their maintainers: as *Ioannes, Accursius Hypolitus* and *Alciatus*? How Iustly and artificially, was *Africanus* reckening made? Proportionating to the Sommes bequeathed, the Contributions of eche part? Namely, for the hundred presently receiued, $17\frac{1}{7}$. And for the hundred, receiued after ten monethes, $12\frac{6}{7}$: which make the 30: which were to be cõtributed by the legataries to the heire.

 For,

For, what proportion, 100 hath to 75: the same hath 17 $\frac{1}{2}$ to 13 $\frac{1}{8}$: Which is Sesquitertia: that is, as 4, to 3. which make 7. Wonderfull many places, in the Ciuile law, require an expert *Arithmeticien*, to vnderstand the deepe Iudgemēt, & Iust determinatiō of the Auncient Romaine Lawmakers. But much more expert ought he to be, who should be hable, to decide with æquitie, the infinite varietie of Cases, which do, or may happen, vnder euery one of those lawes and ordinances Ciuile. Hereby, easely, ye may now coniecture: that in the Canon law: and in the lawes of the Realme (which with vs, beare the chief Authoritie), Iustice and equity might be greately preferred, and skilfully executed, through due skill of Arithmetike, and proportions appertainyng. The worthy Philosophers, and prudent lawmakers (who haue written many bookes *De Republica*: How the best state of Common wealthes might be procured and mainteined,) haue very well determined of Iustice: (which, not onely, is the Base and foundacion of Common weales: but also the totall perfection of all our workes, words, and thoughtes:) defining it, to be that vertue, by which, to euery one, is rendred, that to him appertaineth. (*Iustice.*) God challengeth this at our handes, to be honored as God: to be loued, as a father: to be feared as a Lord & master. Our neighbours proportiō, is also prescribed of the Almighty lawmaker: which is, to do to other, euen as we would be done vnto. These proportions, are in Iustice necessary: in duety, commendable: and of Common wealthes, the life, strength, stay and florishing. *Aristotle* in his *Ethikes* (to fatch the sede of Iustice, and light of direction, to vse and execute the same) was fayne to fly to the perfection, and power of Numbers: for proportions Arithmeticall and Geometricall. *Plato* in his booke called *Epinomis* (which boke, is the Threasury of all his doctrine) where, his purpose is, to seke a Science, which, when a man had it, perfectly: he might seme, and so be, in dede, *Wise*. He, briefly, of other Sciences discoursing, findeth them, not hable to bring it to passe: But of the Science of Numbers, he sayth. *Illa, quæ numerum mortalium generi dedit, id profecto efficiet. Deum autem aliquem, magis quam fortunam, ad salutem nostram, hoc munus nobis arbitror contulisse. &c. Nam ipsum bonorum omnium Authorem, cur non maximi boni, Prudentiæ dico, causam arbitramur? That Science, verely, which hath taught mankynde number, shall be able to bryng it to passe. And, I thinke, a certaine God, rather then fortune, to haue giuen vs this gift, for our blisse. For, why should we not Iudge him, who is the Author of all good things, to be also the cause of the greatest good thyng, namely, Wisedome?* There, at length, he proueth *Wisedome* to be atteyned, by good Skill of *Numbers*. With which great Testimony, and the manifold profes, and reasons, before expressed, you may be sufficiently and fully persuaded: of the perfect Science of *Arithmetike*, to make this accounte: That of all Sciences, next to *Theologie*, it is most diuine, most pure, most ample and generall, most profounde, most subtile, most commodious and most necessary. Whose next Sister, is the Absolute Science of *Magnitudes*: of which (by the Direction and aide of him, whose *Magnitude* is Infinite, and of vs Incomprehensible) I now entend, so to write, that both with the *Multitude*, and also with the *Magnitude* of Meruaylous and frutefull verities, you (my frendes and Countreymen) may be stird vp, and awaked, to behold what certaine Artes and Sciences, (to our vnspeakable behofe) our heauenly father, hath for vs prepared, and reuealed, by sundry *Philosophers* and *Mathematiciens*.

BOth, *Number* and *Magnitude*, haue a certaine Originall sede, (as it were) of an incredible property: and of man, neuer hable, Fully, to be declared. Of *Number*, an Vnit, and of *Magnitude*, a Poynte, doo seeme to be much like Originall

nall

nall causes : But the diuersitie neuerthelesse,is great. We defined an *Vnit*, to be a thing Mathematicall Indiuisible : A Point, likewise, we sayd to be a Mathematicall thing Indiuisible. And farder, that a Point may haue a certaine determined Situation: that is, that we may assigne,and prescribe a Point,to be here, there, yonder. &c. Herein, (behold) our Vnit is free, and can abyde no bondage,or to be tyed to any place,or seat: diuisible or indiuisible. Agayne, by reason,a Point may haue a Situation limited to him: a certaine motion,therfore (to a place,and from a place) is to a Point incident and appertainyng. But an *Vnit*,can not be imagined to haue any motion. A Point,by his motion, produceth, Mathematically,a line: (as we sayd before) which is the first kinde of Magnitudes,and most simple: An *Vnit*,can not produce any number. A Line, though it be produced of a Point moued,yet,it doth not consist of pointes : Number, though it be not produced of an *Vnit*, yet doth it Consist of vnits, as a materiall cause. But formally,Number,is the Vnion, and Vnitie of Vnits. Which vnyting and knitting,is the workemanship of our minde: which,of distinct and discrete Vnits, maketh a Number: by vniformitie,resulting of a certaine multitude of Vnits. And so, euery number,may haue his least part,giuen: namely,an Vnit: But not of a Magnitude, (no,not of a Lyne,) the least part can be giuẽ: bycause,infinitly, diuision therof,may be conceiued. All Magnitude,is either a Line,a Plaine, or a Solid. Which Line,Plaine,or Solid,of no Sense,can be perceiued, nor exactly by hãd (any way) represented: nor of Nature produced: But, as (by degrees) Number did come to our perceiuerance: So,by visible formes,we are holpen to imagine, what our Line Mathematicall, is. What our Point, is. So precise, are our Magnitudes, that one Line is no broader then an other: for they haue no bredth : Nor our Plaines haue any thicknes. Nor yet our Bodies,any weight: be they neuer so large of dimensiõ. Our Bodyes, we can haue Smaller, then either Arte or Nature can produce any : and Greater also, then all the world can comprehend. Our least Magnitudes,can be diuided into so many partes, as the greatest. As, a Line of an inch long, (with vs) may be diuided into as many partes, as may the diameter of the whole world, from East to West : or any way extended : What priuiledges, aboue all manual Arte,and Natures might,haue our two Sciences Mathematicall? to exhibite,and to deale with thinges of such power, liberty, simplicity,puritie,and perfection? And in them,so certainly,so orderly,so precisely to procede: as,excellent is that workemã Mechanicall Iudged, who nerest can approche to the representing of workes, Mathematically demonstrated? And our two Sciences,remaining pure,and absolute,in their proper termes,and in their owne Matter: to haue,and allowe,onely such Demonstrations, as are plaine, certaine, vniuersall, and of an æternall veritye? This Science of *Magnitude*,his properties,conditions,and appertenances : commonly,now is,and from the beginnyng, hath of all Philosophers, ben called *Geometrie*. But,veryly,with a name to base and scant, for a Science of such dignitie and amplenes. And,perchaunce, that name,by cõmon and secret consent,of all wisemen, hitherto hath ben suffred to remayne: that it might carry with it a perpetuall memorye, of the first and notablest benefite, by that Science, to common people shewed : Which was, when Boundes and meres of land and ground were lost, and confounded (as in *Egypt*,yearely,with the ouerflowyng of *Nilus*,the greatest and longest riuer in the world) or, that ground bequeathed,were to be assigned: or, ground sold, were to be layd out : or (when disorder preuailed) that Commõs were distributed into seueralties. For, where, vpon these & such like occasiõs,Some by ignorãce, some by negligẽce, Some by fraude, and some by violence, did wrongfully limite,measure, encroach,or challenge (by

Number.

Geometrie.

 pretence

pretence of iust content, and measure) those landes and groundes : great losse, disquietnes, murder, and warre did (full oft) ensue: Till, by Gods mercy, and mans Industrie, The perfect Science of Lines, Plaines, and Solides (like a diuine Iusticier,) gaue vnto euery man, his owne. The people then, by this art pleasured, and greatly relieued, in their landes iust measuring: & other Philosophers, writing Rules for land measuring: betwene them both, thus, confirmed the name of *Geometria*, that is, (according to the very etimologie of the word) Land measuring. Wherin, the people knew no farder, of Magnitudes vse, but in Plaines: and the Philosophers, of thē, had no feet hearers, or Scholers: farder to disclose vnto, then of flat, plaine *Geometrie*. And though, these Philosophers, knew of farder vse, and best vnderstode the etymologye of the worde, yet this name *Geometria*, was of them applyed generally to all sortes of Magnitudes : vnleast, otherwhile, of *Plato*, and *Pythagoras* : When they would precisely declare their owne doctrine. Then, was * *Geometria*, with them, *Studium quod circa planum versatur*. But, well you may perceiue by *Euclides Elementes*, that more ample is our Science, then to measure Plaines: and nothyng lesse therin is tought (of purpose) then how to measure Land. An other name, therfore, must nedes be had, for our Mathematicall Science of Magnitudes : which regardeth neither clod, nor turff: neither hill, nor dale: neither earth nor heauen: but is absolute *Megethologia*: not creping on ground, and dasseling the eye, with pole » perche, rod or lyne: but liftyng the hart aboue the heauens, by inuisible lines, and » immortall beames meteth with the reflexions, of the light incomprehensible: and » so procureth Ioye, and perfection vnspeakable. Of which true vse of our *Megethica*, or *Megethologia*, *Diuine Plato* seemed to haue good taste, and iudgement: and (by the name of *Geometrie*) so noted it: and warned his Scholers therof: as, in hys seuenth *Dialog*, of the Common wealth, may euidently be sene. Where (in Latin) thus it is: right well translated: *Profecto, nobis hoc non negabunt, Quicunq; vel paululum quid Geometriæ gustârunt, quin hæc Scientia, contrà, omnino se habeat, quàm de ea loquuntur, qui in ipsa versantur*. In English, thus. *Verely* (sayth *Plato*) *whosoeuer haue, (but euen very litle) tasted of Geometrie, will not denye vnto vs, this: but that this Science, is of an other condicion, quite contrary to that, which they that are exercised in it, do speake of it.* And there it followeth, of our *Geometrie*, *Quòd quæritur cognoscendi illius gratia, quod semper est, non & eius quod oritur quandoq; & interit. Geometria, eius quod est semper, Cognitio est. Attollet igitur (ô Generose vir) ad Veritatem, animum: atq; ita, ad Philosophandum preparabit cogitationem, vt ad supera conuertamus: quæ, nunc, contra quàm decet, ad inferiora deijcimus. &c. Quàm maximè igitur præcipiendum est, vt qui præclarissimam hanc habitāt Ciuitatem, nullo modo, Geometriam spernant. Nam & quæ præter ipsius propositum, quodam modo esse videntur, haud exigua sunt. &c.* It must nedes be confessed (saith *Plato*) *That* [*Geometrie*] *is learned, for the knowyng of that, which is euer: and not of that, which, in tyme, both is bred and is brought to an ende. &c. Geometrie is the knowledge of that which is euerlastyng. It will lift vp therfore (O Gentle Syr) our mynde to the Veritie: and by that meanes, it will prepare the Thought, to the Philosophicall loue of wisdome: that we may turne or conuert, toward heauenly thinges* [*both mynde and thought*] *which now, otherwise then becommeth vs, we cast down on base or inferior things. &c. Chiefly, therfore, Commaundement must be giuen, that such as do inhabit this most honorable Citie, by no meanes, despise Geometrie. For euen those thinges* [*done by it*] *which, in manner, seame to be, beside the purpose of Geometrie: are of*

*Plato. 7. de Rep.

☞

no

no ſmall importance. &c. And beſides the manifold vſes of *Geometrie*, in matters appertainyng to warre, he addeth more, of ſecond vnpurpoſed frute, and commoditye, ariſing by *Geometrie*: ſaying: *Scimus quin etiam, ad Diſciplinas omnes facilius perdiſcendas, intereſſe omnino, attigerit ne Geometriam aliquis, an non. &c. Hanc ergo Doctrinam, ſecundo loco diſcendam Iuuenibus ſtatuamus*. That is. *But, alſo, we know, that for the more eaſy learnyng of all Artes, it importeth much, whether one haue any knowledge in Geometrie, or no. &c. Let vs therfore make an ordinance or decree, that this Science, of young men ſhall be learned in the ſecond place.* This was *Diuine Plato* his Iudgement, both of the purpoſed, chief, and perfect vſe of *Geometrie*: and of his ſecond, dependyng, deriuatiue commodities. And for vs, Chriſten men, a thouſand thouſand mo occaſions are, to haue nede of the helpe of* *Megethologicall* Contemplations: wherby, to trayne our Imaginations and Myndes, by litle and litle, to forſake and abandon, the groſſe and corruptible Obiectes, of our vtward ſenſes: and to apprehend, by ſure doctrine demonſtratiue, Things Mathematicall. And by them, readily to be holpen and conducted to conceiue, diſcourſe, and conclude of things Intellectual, Spirituall, æternall, and ſuch as concerne our Bliſſe euerlaſting: which, otherwiſe (without Speciall priuiledge of Illumination, or Reuelation frõ heauen) No mortall mans wyt (naturally) is hable to reach vnto, or to Compaſſe. And, veryly, by my ſmall Talent (from aboue) I am hable to proue and teſtifie, that the litterall Text, and order of our diuine Law, Oracles, and Myſteries, require more ſkill in Numbers, and Magnitudes: then (commonly) the expoſitors haue vttered: but rather onely (at the moſt) ſo warned: & ſhewed their own want therin. (To name any, is nedeles: and to note the places, is, here, no place: But if I be duely aſked, my anſwere is ready.) And without the litterall, Grammaticall, Mathematicall or Naturall verities of ſuch places, by good and certaine Arte, perceiued, no Spirituall ſenſe (propre to thoſe places, by Abſolute *Theologie*) will thereon depend. » No man, therfore, can doute, but toward the atteyning of knowledge incomparable, and Heauenly Wiſedome: Mathematicall Speculations, both of Numbers and Magnitudes: are meanes, aydes, and guides: ready, certaine, and neceſſary. « From henceforth, in this my Preface, will I frame my talke, to *Plato* his fugitiue Scholers: or, rather, to ſuch, who well can, (and alſo wil,) vſe their vtward ſenſes, to the glory of God, the benefite of their Countrey, and their owne ſecret contentation, or honeſt preferment, on this earthly Scaffold. To them, I will orderly recite, deſcribe & declare a great Number of Artes, from our two Mathematicall fountaines, deriued into the fieldes of *Nature*. Wherby, ſuch Sedes, and Rotes, as lye depe hyd in the groũd of *Nature*, are refreſhed, quickened, and prouoked to grow, ſhote vp, floure, and giue frute, infinite, and incredible. And theſe Artes, ſhalbe ſuch, as vpon Magnitudes properties do depende, more, then vpon Number. And by good reaſon we may call them Artes, and Artes Mathematicall Deriuatiue: for (at this tyme) I Define An Arte, to be a Methodicall cõplete Doctrine, hauing abundancy of ſufficient, and peculier matter to deale with, by the allowance of the Metaphiſicall Philoſopher: the knowledge whereof, to humaine ſtate is neceſſarye. And that I account, An Art Mathematicall deriuatiue, which by Mathematicall demonſtratiue Method, in Nũbers, or Magnitudes, ordreth and confirmeth his doctrine, as much & as perfectly, as the matter ſubiect will admit. And for that, I entend

I. D. * *Herein, I would gladly ſhake of, the earthly name, of Geometrie.*

☞

An Arte.

Art Mathematicall Deriuatiue.

A Mechanitien.

I entend to vse the name and propertie of a *Mechanicien*, otherwise, then (hitherto) it hath ben vsed, I thinke it good, (for distinction sake) to giue you also a brief description, what I meane therby. A Mechanicien, or a Mechanicall workman is he, whose skill is, without knowledge of Mathematicall demonstration, perfectly to worke and finishe any sensible worke, by the Mathematicien principall or deriuatiue, demonstrated or demonstrable. Full well I know, that he which inuenteth, or maketh these demonstrations, is generally called *A speculatiue Mechanicien*: which differreth nothyng from a *Mechanicall Mathematicien*. So, in respect of diuerse actions, one man may haue the name of sundry artes: as, some tyme, of a Logicien, some tymes (in the same matter otherwise handled) of a Rethoricien. Of these trifles, I make, (as now, in respect of my Preface,) small account: to fyle thẽ for the fine handlyng of subtile curious disputers. In other places, they may commaunde me, to giue good reason: and yet, here, I will not be vnreasonable.

Geometrie vulgar.

First, then, from the puritie, absolutenes, and Immaterialitie of Principall *Geometrie*, is that kinde of *Geometrie* deriued, which vulgarly is counted *Geometrie*: and is the Arte of Measuring sensible magnitudes, their iust quãtities and contentes. This, teacheth to measure, either at hand: and the practiser, to be by the thing Measured: and so, by due applying of Cumpase, Rule, Squire, Yarde, Ell, Perch, Pole, Line, Gaging rod, (or such like instrument) to the Length, Plaine, or Solide measured, *to be certified, either of the length, perimetry, or distance lineall: and this is called, *Mecometrie*. Or *to be certified of the content of any plaine Superficies: whether it be in ground Surueyed, Borde, or Glasse measured, or such like thing: which measuring, is named *Embadometrie*. *Or els to vnderstand the Soliditie, and content of any bodily thing: as of Tymber and Stone, or the content of Pits, Pondes, Wells, Vessels, small & great, of all fashions. Where, of Wine, Oyle, Beere, or Ale vessells, &c, the Measuring, commonly, hath a peculier name: and is called *Gaging*. And the generall name of these Solide measures, is *Stereometrie*. Or els, this *vulgar Geometrie*, hath consideration to teach the practiser, how to measure things, with good distance betwene him and the thing measured: and to vnderstand thereby, either *how Farre, a thing seene (on land or water) is from the measurer: and this may be called *Apomecometrie*: Or, how High or depe, aboue or vnder the leuel of the measurers stãding, any thing is, which is sene on land or water, called *Hypsometrie*. *Or, it informeth the measurer, how Broad any thing is, which is in the measurers vew: so it be on Land or Water, situated: and may be called *Platometrie*. Though I vse here to condition, the thing measured, to be on Land, or Water Situated: yet, know for certaine, that the sundry heigthe of Cloudes, blasing Starres, and of the Mone, may (by these meanes) haue their distances from the earth: and, of the blasing Starres and Mone, the Soliditie (aswell as distances) to be measured: But because, neither these things are vulgarly taught: nor of a common practiser so ready to be executed: I, rather, let such measures be reckened incident to some of our other Artes, dealing with thinges on high, more purposely, then this vulgar Land measuring Geometrie doth: as in *Perspectiue* and *Astronomie, &c*.

Note.

(Margin numbers: I. 1. 2. 3. 2. 1. 2. 3.)

OF these Feates (farther applied) is Sprong the Feate of *Geodesie*, or Land Measuring: more cunningly to measure & Suruey Land, Woods, and Waters, a farre of. More cunningly, I say: But God knoweth (hitherto) in these Realmes of England and Ireland (whether through ignorance or fraude, I can not tell, in euery particular) how great wrong and iniurie hath (in my time) bene committed

Note.

by

by vntrue measuring and surueying of Land or Woods, any way. And, this I am sure: that the Value of the difference, betwene the truth and such Surueyes, would haue bene hable to haue foũd (for euer) in eche of our two Vniuersities, an excellent Mathematicall Reader: to eche, allowing (yearly) a hundred Markes of lawfull money of this realme: which, in dede, would seme requisit, here, to be had (though by other wayes prouided for) as well, as, the famous Vniuersitie of Paris, hath two Mathematicall Readers: and eche, two hundreth French Crownes yearly, of the French Kinges magnificent liberalitie onely. Now, againe, to our purpose returning: Moreouer, of the former knowledge Geometricall, are growen the Skills of *Geographie*, *Chorographie*, *Hydrographie*, and *Stratarithmetrie*.

Geographie teacheth wayes, by which, in sũdry formes, (as *Sphærike, Plaine* ,, or other), the Situation of Cities, Townes, Villages, Fortes, Castells, Mountaines, ,, Woods, Hauens, Riuers, Crekes, & such other things, vpõ the outface of the earthly ,, Globe (either in the whole, or in some principall mẽber and portion therof cõtayned) ,, may be described and designed, in cõmensurations Analogicall to Nature ,, and veritie: and most aptly to our vew, may be represented. Of this Arte how great ,, pleasure, and how manifolde commodities do come vnto vs, daily and hourely: of most men, is perceaued. While, some, to beautifie their Halls, Parlers, Chambers, Galeries, Studies, or Libraries with: other some, for thinges past, as battels fought, earthquakes, heauenly fyringes, & such occurentes, in histories mentioned: therby liuely, as it were, to vewe the place, the region adioyning, the distance from vs: and such other circumstances. Some other, presently to vewe the large dominion of the Turke: the wide Empire of the Moschouite: and the litle morsell of ground, where Christendome (by profession) is certainly knowen. Litle, I say, in respecte of the rest. &c. Some, either for their owne iorneyes directing into farre landes: or to vnderstand of other mens trauailes. To conclude, some, for one purpose: and some, for an other, liketh, loueth, getteth, and vseth, Mappes, Chartes, & Geographicall Globes. Of whose vse, to speake sufficiently, would require a booke peculier.

Chorographie seemeth to be an vnderling, and a twig, of *Geographie*: and yet neuerthelesse, is in practise manifolde, and in vse very ample. This tea- ,, cheth Analogically to describe a small portion or circuite of ground, with the con- ,, tentes: not regarding what commensuration it hath to the whole, or any parcell, ,, without it, contained. But in the territory or parcell of ground which it taketh in ,, hand to make description of, it leaueth out (or vndescribed) no notable, or odde ,, thing, aboue the ground visible. Yea and sometimes, of thinges vnder ground, ,, geueth some peculier marke: or warning: as of Mettall mines, Cole pittes, Stone ,, quarries. &c. Thus, a Dukedome, a Shiere, a Lordship, or lesse, may be described ,, distinctly. But marueilous pleasant, and profitable it is, in the exhibiting to our eye, and commensuration, the plat of a Citie, Towne, Forte, or Pallace, in true Symmetry: not approching to any of them: and out of Gunne shot. &c. Hereby, the *Architect* may furnishe him selfe, with store of what patterns he liketh: to his great instruction: euen in those thinges which outwardly are proportioned: either simply in them selues: or respectiuely, to Hilles, Riuers, Hauens, and Woods adioyning. Some also, terme this particular description of places, *Topographie*.

Hydrographie, deliuereth to our knowledge, on Globe or in Plaine, ,, the perfect Analogicall description of the Ocean Sea coastes, through the whole ,, world: or in the chiefe and principall partes thereof: with the Iles and chiefe ,, particular places of daungers, conteyned within the boundes, and Sea coastes described: as, of Quickfandes, Bankes, Pittes, Rockes, Races, Countertides, Whorlepooles. &c. This, dealeth with the Element of the water chiefly: as *Geographie* did principally take the Element of the Earthes description (with his appertenances) to taske. And besides thys, *Hydrographie*, requireth a particular Register of certaine Landmarkes (where markes may be had) from the sea, well hable to be skried, in what point of the Seacumpase they appeare, and what apparent forme, Situation, and bignes they haue, in respecte of any daungerous place in the sea, or nere vnto it, assigned: And in all Coastes, what Mone, maketh full Sea: and what way, the Tides and Ebbes, come and go, the *Hydrographer* ought to recorde. The Soundinges likewise: and the Chanels wayes: their number, and depthes ordinarily, at ebbe and flud, ought the *Hydrographer*, by obseruation and diligence of *Measuring*, to haue certainly knowen. And many other pointes, are belonging to perfecte *Hydrographie*, and for to make a *Rutter*, by: of which, I nede not here speake: as of the describing, in any place, vpon Globe or Plaine, the 32. pointes of the Compase, truely: (wherof, scarsly foure, in England, haue right knowledge: bycause, the lines therof, are no straight lines, nor Circles.) Of making due proiection of a Sphere in plaine. Of the Variacion of the Compas, from true Northe: And such like matters (of great importance, all) I leaue to speake of, in this place: bycause, I may seame (al ready) to haue enlarged the boundes, and duety of an Hydographer, much more, then any man (to this day) hath noted, or prescribed. Yet am I well hable to proue, all these thinges, to appertaine, and also to be proper to the Hydrographer. The chief vse and ende of this Art, is the Art of Nauigation: but it hath other diuerse vses: euen by them to be enioyed, that neuer lacke sight of land.

Stratarithmetrie, is the Skill, (appertainyng to the warre,) by which a man can set in figure, analogicall to any *Geometricall* figure appointed, any certaine number or summe of men: of such a figure capable: (by reason of the vsuall spaces betwene Souldiers allowed: and for that, of men, can be made no Fractions. Yet, neuertheles, he can order the giuen summe of men, for the greatest such figure, that of them, cā be ordred) and certifie, of the ouerplus: (if any be) and of the next certaine summe, which, with the ouerplus, will admit a figure exactly proportionall to the figure assigned. By which Skill, also, of any army or company of men: (the figure & sides of whose orderly standing, or array, is knowen) he is able to expresse the iust number of men, within that figure conteined: or (orderly) able to be conteined. * And this figure, and sides therof, he is hable to know: either beyng by, and at hand: or a farre of. Thus farre, stretcheth the description and property of *Stratarithmetrie*: sufficient for this tyme and place. It differreth from the Feate *Tacticall, De aciebus instruendis*. bycause, there, is necessary the wisedome and foresight, to what purpose he so ordreth the men: and Skillfull hability, also, for any occasion, or purpose, to deuise and vse the aptest and most necessary order, array and figure of his Company and Summe of men. By figure, I meane: as, either of a *Perfect Square, Triangle, Circle, Ouale, long square,* (of the Grekes it is called *Eteromekes*) *Rhombe, Rhomboid, Lunular, Ryng, Serpentine,* and such other Geometricall figures: Which, in warres, haue ben, and are to be vsed: for commodiousnes, necessity, and auauntage &c. And no small skill ought he to haue, that should make true report, or nere the truth, of the numbers and Summes, of footemen or horsemen, in the Enemyes ordring. A farre of, to make an estimate, betwene nere termes of More and Lesse, is not a thyng very rise, among those that gladly would do

*Note.

The difference betwene Stratarithmetrie and Tacticie.

do it. Great pollicy may be vsed of the Capitaines,(at tymes fete, and in places conuenient)as to vse Figures, which make greatest shew, of so many as he hath: and vsing the aduauntage of the three kindes of vsuall spaces: (betwene footemen or horsemen)to take the largest: or when he would seme to haue few, (beyng many:)contrary wise,in Figure,and space. The Herald,Pursuant, Sergeant Royall, Capitaine, or who soeuer is carefull to come nere the truth herein, besides the Iudgement of his expert eye,his skill of Ordering *Tacticall*, the helpe of his Geometricall instrument: Ring, or Staffe Astronomicall: (commodiously framed for cariage and vse) He may wonderfully helpe him selfe, by perspectiue Glasses. In which, (I trust) our posterity will proue more skillfull and expert, and to greater purposes, then in these dayes, can(almost)be credited to be possible.

I.D. Frende, you will finde it hard, to performe my description of this Feate. But by Chorographie, you may helpe your selfe some what: where the Figures knowne (in Sides and Angles) are not Regular: And where, Resolution into Triangles can serue. &c. And yet you will finde it strange to deale thus generally with Arithmeticall figures: and, that for Batayle ray. Their contentes, differ so much from like Geometricall Figures.

Thus haue I lightly passed ouer the Artificiall Feates, chiefly dependyng vpon vulgar *Geometrie*: & commonly and generally reckened vnder the name of *Geometrie*. But there are other(very many) *Methodicall Artes*, which, declyning from the purity, simplicitie, and Immateriality, of our Principall Science of *Magnitudes*: do yet neuertheles vse the great ayde, direction, and Method of the sayd principall Science, and haue propre names, and distinct: both from the Science of *Geometrie*, (from which they are deriued)and one from the other. As Perspectiue, Astronomie, Musike, Cosmographie, Astrologie, Statike, Anthropographie, Trochilike, Helicosophie, Pneumatithmie, Menadrie, Hypogeiodie, Hydragogie, Horometrie, Zographie, Architecture, Nauigation, Thaumaturgike and Archemastrie. I thinke it necessary, orderly, of these to giue some peculier descriptions: and withall, to touch some of their commodious vses, and so to make this Preface, to be a little swete, pleasant Nosegaye for you: to comfort your Spirites, beyng almost out of courage, and in despayre, (through brutish brute) Weenyng that *Geometrie*, had but serued for buildyng of an house, or a curious bridge, or the roufe of Westminster hall, or some witty pretty deuise, or engyn, appropriate to a Carpenter, or a Ioyner &c. That the thing is farre otherwise, then the world, (commonly) to this day, hath demed, by worde and worke, good profe wilbe made.

Among these Artes, by good reason, Perspectiue ought to be had, ere of *Astronomicall Apparences*, perfect knowledge can be atteyned. And bycause of the prerogatiue of *Light*, beyng the first of *Gods Creatures*: and the eye, the light of our body, and his Sense most mighty, and his organ most Artificiall and *Geometricall*: At *Perspectiue*, we will begyn therfore. Perspectiue, is an Art Mathematicall, which demonstrateth the maner, and properties, of all Radiations Direct, Broken, and Reflected. This Description, or Notation, is brief: but it reacheth so farre, as the world is wyde. It concerneth all Creatures, all Actions, and passions, by Emanation of beames perfourmed. Beames, or naturall lines, (here) I meane, not of light onely, or of colour (though they, to eye, giue shew, witnes, and profe, wherby to ground the Arte vpon) but also of other *Formes*, both *Substantiall*, and *Accidentall*, the certaine and determined actiue Radiall emanations. By this Art (omitting to speake of the highest pointes) we may vse our eyes, and the light, with greater pleasure: and perfecter Iudgement: both of things, in light seen, & of other: which by like order of Lightes Radiations, worke and produce their effectes. We may be ashamed to be ignorant of the cause, why so sundry wayes our eye is deceiued, and abused: as, while the eye weeneth a roũd Globe or Sphere (beyng farre of) to be a flat and plaine Circle, and so likewise iudgeth a plaine Square, to be roũd: ſuppoſeth walles parallels, to approche, a farre of: rofe and floure parallels, the one to bend downward, the other to riſe vpward, at a little diſtance from you. Againe, of thinges being in like ſwiftnes of mouing, to thinke the nerer, to moue faſter: and the farder, much ſlower. Nay, of two thinges, wherof the one (incomparably) doth moue ſwifter then the other, to deme the ſlower to moue very ſwift, & the other to ſtand: what an error is this, of our eye? Of the Raynbow, both of his Colours, of the order of the colours, of the bignes of it, the place and heith of it, (&c) to know the cauſes demonſtratiue, is it not pleaſant, is it not neceſſary? of two or three Sonnes appearing: of Blaſing Sterres: and ſuch like thinges: by naturall cauſes, brought to paſſe, (and yet neuertheles, of farder matter, Significatiue) is it not commodious for man to know the very true cauſe, & occaſion Naturall? Yea, rather, is it not, greatly, againſt the Soueraintу of Mans nature, to be ſo ouerſhot and abuſed, with thinges (at hand) before his eyes? as with a Pecockes tayle, and a Doues necke: or a whole ore, in water, holden, to ſeme broken. Thynges, farre of, to ſeeme nere: and nere, to ſeme farre of. Small thinges, to ſeme great: and great, to ſeme ſmall. One man, to ſeme an Army. Or a man to be curſtly affrayed of his owne ſhaddow. Yea, ſo much, to feare, that, if you, being (alone) nere a certaine glaſſe, and proffer, with dagger or ſword, to foyne at the glaſſe, you ſhall ſuddenly be moued to giue backe (in maner) by reaſon of an Image, appearing in the ayre, betwene you & the glaſſe, with like hand, ſword or dagger, & with like quicknes, foyning at your very eye, likewiſe as you do at the Glaſſe. Straunge, this is, to heare of: but more meruailous to behold, then theſe my wordes can ſignifie. And neuertheleſſe by demonſtration Opticall, the order and cauſe therof, is certified: euen ſo, as the effect is conſequent. Yea, thus much more, dare I take vpon me, toward the ſatiſfying of the noble courrage, that longeth ardently for the wiſedome of Cauſes Naturall: as to let him vnderſtand, that, in London, he may with his owne eyes, haue profe of that, which I haue ſayd herein. A Gentleman, (which, for his good ſeruice, done to his Countrey, is famous and honorable: and for ſkill in the Mathematicall Sciences, and Languages, is the Od man of this land. &c.) euen he, is hable: and (I am ſure) will, very willingly, let the Glaſſe, and profe be ſene: and ſo I (here) requeſt him: for the encreaſe of wiſedome, in the honorable: and for the ſtopping of the mouthes malicious: and repreſſing the arrogancy of the ignorant. Ye may eaſily geſſe, what I meane. This Art of *Perſpectiue*, is of that excellency, and may be led, to the certifying, and executing of ſuch thinges, as no man would eaſily beleue: without Actuall profe perceiued. I ſpeake nothing of *Naturall Philoſophie*, which, without *Perſpectiue*, can not be fully vnderſtanded, nor perfectly atteined vnto. Nor, of *Aſtronomie*: which, without *Perſpectiue*, can not well be grounded: Nor *Aſtrologie*, naturally Verified, and auouched. That part hereof, which dealeth with Glaſſes (which name, Glaſſe, is a generall name, in this Arte, for any thing, from which, a Beame reboundeth) is called *Catoptrike*: and hath ſo many vſes, both merueilous, and proffitable: that, both, it would hold me to long, to note therin the principall concluſions, all ready knowne: And alſo (perchaunce) ſome thinges, might lacke due credite with you: And I, therby, to leeſe my labor: and you, to ſlip into light Iudgement*, Before you haue learned ſufficiently the powre of Nature and Arte.

A marueilous Glaſſe. ☞

S.W.P.

☞

Now, to procede: Aſtronomie, is an Arte Mathematicall, which demonſtrateth the diſtance, magnitudes, and all naturall motions, apparences, and paſsions propre to the Planets and fixed Sterres: for

any

any time paſt,preſent and to come: in reſpect of a certaine Horizon, or without reſpect of any Horizon. By this Arte we are certified of the diſtance of the Starry Skye,and of eche *Planete* from the Centre of the Earth: and of the greatnes of any Fixed ſtarre ſene, or *Planete*,in reſpect of the Earthes greatnes. As, we are ſure (by this Arte) that the Solidity, Maſſines and Body of the *Sonne*, conteineth the quantitie of the whole Earth and Sea,a hundred thre ſcore and two times, leſſe by $\frac{1}{8}$ one eight parte of the earth. But the Body of the whole earthly globe and Sea,is bigger then the body of the Mone, three and forty times leſſe by $\frac{1}{8}$ of the Mone. Wherfore the *Sonne* is bigger then the *Mone*, 7000 times, leſſe, by 59 $\frac{19}{64}$ that is, preciſely 6940 $\frac{21}{64}$ bigger then the *Mone*. And yet the vnſkillfull man,would iudge them a like bigge. Wherfore,of Neceſsity,the one is much farder from vs,then the other. The *Sonne*, when he is fardeſt from the earth (which,now,in our age,is,when he is in the 8.degree,of Cancer) is, 1179. Semidiameters of the Earth,diſtante. And the *Mone* when ſhe is fardeſt from the earth,is 68 Semidiameters of the earth and $\frac{1}{3}$ The nereſt, that the *Mone* commeth to the earth,is Semidiameters 52 $\frac{1}{4}$ The diſtance of the Starry Skye is, frõ vs,in Semidiameters of the earth 20081 $\frac{1}{2}$ Twenty thouſand fourescore, one, and almoſt a halfe. Subtract from this,the *Mones* nereſt diſtance,from the Earth: and therof remaineth Semidiameters of the earth 20029 $\frac{1}{4}$ Twenty thouſand nine and twenty and a quarter. So thicke is the heauenly Palace, that the *Planetes* haue all their exerciſe in,and moſt meruailouſly perfourme the Commaũdement and Charge to them giuen by the omnipotent Maieſtie of the king of kings. This is that, which in *Geneſis* is called *Ha Rakia*. Conſider it well. The Semidiameter of the earth, cõteineth of our common miles 3436 $\frac{4}{11}$ three thouſand,foure hundred thirty ſix and foure eleuenth partes of one myle: Such as the whole earth and Sea, round about, is 21600. One and twenty thouſand ſix hundred of our myles. Allowyng for euery degree of the greateſt circle,thre ſcore myles. Now if you way well with your ſelfe but this litle parcell of frute *Aſtronomicall*, as concerning the bigneſſe,Diſtances of *Sonne,Mone, Sterry Sky*,and the huge maſſines of *Ha Rakia*, will you not finde your Conſciences moued, with the kingly Prophet, to ſing the confeſſion of Gods Glory,and ſay, *The Heauens declare the glory of God,and the Firmament* [*Ha Rakia*] *ſheweth forth the workes of his handes.* And ſo forth,for thoſe fiue firſt ſtaues,of that kingly Pſalme. Well,well,It is time for ſome to lay hold on wiſedome,and to Iudge truly of thinges: and not ſo to expound the Holy word,all by Allegories: as to Neglect the wiſedome, powre and Goodnes of God,in, and by his Creatures, and Creation to be ſeen and learned. By parables and Analogies of whoſe natures and properties,the courſe of the Holy Scripture,alſo, declareth to vs very many Myſteries. The whole Frame of Gods Creatures,(which is the whole world,) is to vs,a bright glaſſe: from which, by reflexion,reboundeth to our knowledge and perceiuerance, Beames, and Radiations: repreſenting the Image of his Infinite goodnes, Omnipotẽcy,and wiſedome. And we therby, are taught and perſuaded to Glorifie our Creator,as God: and be thankefull therfore. Could the Heatheniſtes finde theſe vſes,of theſe moſt pure, beawtifull,and Mighty Corporall Creatures: and ſhall we,after that the true *Sonne* of rightwiſeneſſe is riſen aboue the *Horizon*,of our temporall *Hemiſpharie*,and hath ſo abundantly ſtreamed into our hartes,the direct beames of his goodnes, mercy, and grace: Whoſe heat All Creatures feele: Spirituall and Corporall: Viſible and

Note.

Inuisible. Shall we (I say) looke vpon the *Heauen, Sterres*, and *Planets*, as an Oxe and an Asse doth: no furder carefull or inquisitiue, what they are: why were they Created, How do they execute that they were Created for? Seing, All Creatures, were for our sake created: and both we, and they, Created, chiefly to glorifie the Almighty Creator: and that, by all meanes, to vs possible. *Nolite ignorare* (saith *Plato in Epinomis*) *Astronomiam, Sapientissimũ quiddam esse. Be ye not ignorant, Astronomie to be a thyng of excellent wisedome.* *Astronomie*, was to vs, from the beginning commended, and in maner commaunded by God him selfe. In asmuch as he made the *Sonne*, *Mone*, and *Sterres*, to be to vs, for *Signes*, and knowledge of Seasons, and for Distinctions of Dayes, and yeares. Many wordes nede not. But I wish, euery man should way this word, *Signes*. And besides that, conferre it also with the tenth Chapter of *Hieremie*. And though Some thinke, that there, they haue found a rod: Yet Modest Reason, will be indifferent Iudge, who ought to be beaten therwith, in respect of our purpose. Leauing that: I pray you vnderstand this: that without great diligence of Obseruation, examination and Calculation, their periods and courses (wherby *Distinction* of Seasons, yeares, and New Mones might precisely be knowne) could not exactely be certified. Which thing to performe, is that *Art*, which we here haue Defined to be *Astronomie*. Wherby, we may haue the distinct Course of Times, dayes, yeares, and Ages: aswell for Consideratiõ of Sacred Prophesies, accomplished in due time, foretold: as for high Mysticall Solemnities holding: And for all other humaine affaires, Conditions, and couenantes, vpon certaine time, betwene man and man: with many other great vses: Wherin, (verely), would be great incertainty, Confusion, vntruth, and brutish Barbarousnes: without the wonderfull diligence and skill of this Arte: continually learning, and determining Times, and periodes of Time, by the Record of the heauenly booke, wherin all times are written: and to be read with an *Astronomicall staffe*, in stede of a festue.

Musike, of Motion, hath his Originall cause: Therfore, after the motions most swift, and most Slow, which are in the Firmament, of Nature performed: and vnder the *Astronomers Consideration*: now I will Speake of an other kinde of *Motion*, producing sound, audible, and of Man numerable. *Musike* I call here that *Science*, which of the Grekes is called *Harmonice*. Not medling with the Controuersie betwene the auncient *Harmonistes*, and *Canonistes*. Musike is a Mathematicall Science, which teacheth, by sense and reason, perfectly to iudge, and order the diuersities of soundes, hye and low. *Astronomie* and *Musike* are Sisters, saith *Plato*. As, for *Astronomie*, the eyes: So, for *Harmonious Motion*, the eares were made. But as *Astronomie* hath a more diuine Contemplation, and cõmodity, then mortall eye can perceiue: So, is *Musike* to be considered, that the

1\. * Minde may be preferred, before the eare. And from audible sound, we ought to ascende, to the examination: which numbers are *Harmonious*, and which not. And why, either, the one are: or the other are not. I could at large, in the heauenly

2\. * motions and distances, describe a meruailous Harmonie, of *Pythagoras* Harpe

4\. 3. with eight stringes. Also, somwhat might be sayd of *Mercurius* * two Harpes, eche of foure Stringes Elementall. And very straunge matter, might be alledged

5\. of the *Harmonie*, to our * Spirituall part appropriate. As in *Ptolomaus* third boke, in the fourth and sixth Chapters may appeare. * And what is the cause of the apt

6\. bonde, and frendly felowship, of the Intellectuall and Mentall part of vs, with our grosse & corruptible body: but a certaine Meane, and *Harmonious Spiritualitie*, *with*

both

both participatyng,& of both (in a maner) resultyng: In the * *Tune of Mans voyce, and also* * *the sound of Instrument*, what might be sayd, of *Harmonie:* No common Musicien would lightly beleue. But of the sundry Mixture (as I may terme it) and concurse, diuerse collation, and Application of these *Harmonies:* as of thre, foure, fiue, or mo: Maruailous haue the effectes ben: and yet may be founde, and produced the like: with some proportionall consideration for our time, and being: in respect of the State, of the thinges then: in which, and by which, the wondrous effectes were wrought. *Democritus* and *Theophrastus* affirmed, that, by *Musike*, griefes and diseases of the Minde, and body might be cured, or inferred. And we finde in Recorde, that *Terpander, Arion, Ismenias, Orpheus, Amphion, Dauid, Pythagoras, Empedocles, Asclepiades* and *Timotheus*, by *Harmonicall* Consonācy, haue done, and brought to pas, thinges, more then meruailous, to here of. Of them then, making no farder discourse, in this place: Sure I am, that Common *Musike*, commonly vsed, is found to the *Musiciens* and Hearers, to be so Commodious and pleasant, That if I would say and dispute, but thus much: That it were to be otherwise vsed, then it is, I should finde more repreeuers, then I could finde priuy, or skilfull of my meaning. In thinges therfore euident, and better knowen, then I can expresse: and so allowed and liked of, (as I would wish, some other thinges, had the like hap) I will spare to enlarge my lines any farder, but consequently follow my purpose.

7.
8.

I.D. Read in Aristotle his 8. booke of Politikes: the 5, 6, and 7. chapters. Where you shall haue some occasion farder to thinke of Musike, than commonly is thought.

Of Cosmographie, I appointed briefly in this place, to geue you some intelligence. Cosmographie, is the whole and perfect description of the heauenly, and also elementall parte of the world, and their homologall application, and mutuall collation necessarie. This Art, requireth *Astronomie, Geographie, Hydrographie* and *Musike*. Therfore, it is no small Arte, nor so simple, as in common practise, it is (slightly) considered. This matcheth Heauen, and the Earth, in one frame, and aptly applieth parts Correspōdent: So, as, the Heauenly Globe, may (in practise) be duely described vpon the Geographicall, and Hydrographicall Globe. And there, for vs to consider an *Æquinoctiall Circle, an Ecliptike line, Colures, Poles, Sterres* in their true Longitudes, Latitudes, Declinations, and Verticalitie: also Climes, and Parallels: and by an *Horizon* annexed, and reuolution of the earthly Globe (as the Heauen, is, by the *Primouant*, caried about in 24. æquall Houres) to learne the Risinges and Settinges of Sterres (of *Virgill* in his *Georgikes*: of *Hesiod*: of *Hippocrates* in his *Medicinall Sphære*, to Perdicca King of the Macedonians: of *Diocles*, to King *Antigonus*, and of other famous *Philosophers* prescribed) a thing necessary, for due manuring of the earth, for *Nauigation*, for the Alteration of mans body: being, whole, Sicke, wounded, or brused. By the Reuolution, also, or mouing of the Globe Cosmographicall, the Rising and Setting of the Sonne: the Lengthes, of dayes and nightes: the Houres and times (both night and day) are knowne: with very many other pleasant and necessary vses: Wherof, some are knowne: but better remaine, for such to know and vse: who of a sparke of true fire, can make a wonderfull bonfire, by applying of due matter, duely.

Of Astrologie, here I make an Arte, seuerall from *Astronomie:* not by new deuise, but by good reason and authoritie: for, Astrologie, is an Arte Mathematicall, which reasonably demonstrateth the operations and effectes, of the naturall beames, of light, and secrete influence: of the Sterres and Planets: in euery element and elementall body:

at all times, in any Horizon assigned. This Arte is furnished with many other great Artes and experiences: As with perfecte *Perspectiue*, *Astronomie*, *Cosmographie*, *Naturall Philosophie* of the 4. Elementes, the Arte of Graduation, and some good vnderstãding in *Musike*: and yet moreouer, with an other great Arte, hereafter following, though I, here, set this before, for some considerations me mouing. Sufficient (you see) is the stuffe, to make this rare and secrete Arte, of: and hard enough to frame to the Conclusion Syllogisticall. Yet both the manifolde and continuall trauailes of the most auncient and wise Philosophers, for the atteyning of this Arte: and by examples of effectes, to confirme the same: hath left vnto vs sufficient proufe and witnesse: and we, also, daily may perceaue, That mans body, and all other Elementall bodies, are altered, disposed, ordred, pleasured, and displeasured, by the Influentiall working of the *Sunne*, *Mone*, and the other Starres and Planets. And therfore, sayth *Aristotle*, in the first of his *Meteorologicall* bookes, in the second Chapter: *Est autem necessariò Mundus iste, supernis lationibus ferè continuus. Vt, inde, vis eius vniuersa regatur. Ea siquidem Causa prima putanda omnibus est, vnde motus principium existit.* That is: *This* [*Elementall*] *World is of necessitie, almost, next adioyning, to the heauenly motions: That, from thence, all his vertue or force may be gouerned. For, that is to be thought the first Cause vnto all: from which, the beginning of motion, is.* And againe, in the tenth Chapter. *Oportet igitur & horum principia sumamus, & causas omnium similiter. Principium igitur vt mouens, præcipuumq́, & omnium primum, Circulus ille est, in quo manifeste Solis latio, &c.* And so forth. His *Meteorologicall* bookes, are full of argumentes, and effectuall demonstrations, of the vertue, operation, and power of the heauenly bodies, in and vpon the fower Elementes, and other bodies, of them (either perfectly, or vnperfectly) composed. And in his second booke, *De Generatione & Corruptione*, in the tenth Chapter. *Quocirca & prima latio, Ortus & Interitus causa non est: Sed obliqui Circuli latio: ea namq́, & continua est, & duobus motibus fit:* In Englishe, thus. *Wherefore the vppermost motion, is not the cause of Generation and Corruption, but the motion of the Zodiake: for, that, both, is continuall, and is caused of two mouinges.* And in his second booke, and second Chapter of hys *Physikes*. *Homo namq́, generat hominem, atq́, Sol. For Man* (sayth he) *and the Sonne, are cause of mans generation.* Authorities may be brought, very many: both of 1000. 2000. yea and 3000. yeares Antiquitie: of great *Philosophers*, *Expert*, *Wise*, and godly men, for that Conclusion: which, daily and hourely, we men, may discerne and perceaue by sense and reason: All beastes do feele, and simply shew, by their actions and passions, outward and inward: All Plants, Herbes, Trees, Flowers, and Fruites. And finally, the Elementes, and all thinges of the Elementes composed, do geue Testimonie (as *Aristotle* sayd) that theyr *Whole Dispositions, vertues, and naturall motions, depend of the Actiuitie of the heauenly motions and Influences. Whereby, beside the specificall order and forme, due to euery seede: and beside the Nature, propre to the Indiuiduall Matrix, of the thing produced: What shall be the heauenly Impression, the perfect and circumspecte Astrologien hath to Conclude.* Not onely (by *Apotelesmes*) τὸ ὅτι, but by Naturall and Mathematicall demonstration τὸ διότι. Whereunto, what Sciences are requisite (without exception) I partly haue here warned: And in my *Propædeumes* (besides other matter there disclosed) I haue Mathematically furnished vp the whole Method: To this our age, not so carefully handled by any, that

euer

euer I saw, or heard of. I was, (for * 21.yeares ago) by certaine earnest disputations, of the Learned *Gerardus Mercator*, and *Antonius Gogaua*, (and other,) therto so prouoked: and (by my constant and inuincible zeale to the veritie) in obseruations of Heauenly Influencies (to the Minute of time,) than, so diligent: And chiefly by the Supernaturall influence, from the Starre of Iacob, so directed: That any Modest and Sober Student, carefully and diligently seking for the Truth, will both finde & cõfesse, therin, to be the Veritie, of these my wordes: And also become a Reasonable Reformer, of three Sortes of people: about these Influentiall Operations, greatly erring from the truth. Wherof, the one, is Light Beleuers, the other, Light Despisers, and the third Light Practisers. The first, & most cõmon Sort, thinke the Heauen and Sterres, to be answerable to any their doutes or desires: which is not so: and, in dede, they, to much, ouer reache. The Second sorte thinke no Influentiall vertue (frõ the heauenly bodies) to beare any Sway in Generation and Corruption, in this Elementall world. And to the *Sunne*, *Mone* and *Sterres* (being so many, so pure, so bright, so wonderfull bigge, so farre in distance, so manifold in their motions, so constant in their periodes. &c.) they assigne a sleight, simple office or two, and so allow vnto thẽ (according to their capacities) as much vertue, and power Influentiall, as to the Signe of the *Sunne*, *Mone*, and seuen Sterres, hanged vp (for Signes) in London, for distinction of houses, & such grosse helpes, in our wordly affaires: And they vnderstand not (or will not vnderstand) of the other workinges, and vertues of the Heauenly *Sunne*, *Mone*, and *Sterres*: not so much, as the Mariner, or Husband man: no, not so much, as the *Elephant* doth, as the *Cynocephalus*, as the Porpentine doth: nor will allow these perfect, and incorruptible mighty bodies, so much vertuall Radiation, & Force, as they see in a litle peece of a *Magnes stone*: which, at great distance, sheweth his operation. And perchaunce they thinke, the Sea & Riuers (as the Thames) to be some quicke thing, and so to ebbe, and flow, run in and out, of them selues, at their owne fantasies. God helpe, God helpe. Surely, these men, come to short: and either are to dull: or willfully blind: or, perhaps, to malicious. The third man, is the common and vulgare *Astrologien*, or Practiser: who, being not duely, artificially, and perfectly furnished: yet, either for vaine glory, or gayne: or like a simple dolt, & blinde Bayard, both in matter and maner, erreth: to the discredit of the *Wary*, and modest *Astrologien*: and to the robbing of those most noble corporall Creatures, of their Naturall Vertue: being most mighty: most beneficiall to all elementall Generation, Corruption and the appartenances: and most Harmonious in their Monarchie: For which thinges, being knowen, and modestly vsed: we might highly, and continually glorifie God, with the princely Prophet, saying. *The Heauens declare the Glorie of God: who made the Heauẽs in his wisedome: who made the Sonne, for to haue dominion of the day: the Mone and Sterres to haue dominion of the nyght: whereby, Day to day vttereth talke: and night, to night declareth knowledge. Prayse him, all ye Sterres, and Light. Amen.*

*Anno. 1548 and 1549. in Louayn.

Note.

1.

2.

3.

IN order, now foloweth, of Statike, somewhat to say, what we meane by that name: and what commodity, doth, on such Art, depend. Statike, is an Arte Mathematicall, which demonstrateth the causes of heauynes, and lightnes of all thynges: and of motions and properties, to heauynes and lightnes, belonging. And for asmuch as, by the Bilanx, or Balance (as the chief sensible Instrument,) Experience of these demonstrations may be had: we call this Art, *Statike:* that is, *the Experimentes of the Balance.* Oh, that men wist, what proffit, (all maner of wayes) by this Arte might grow, to the hable examiner, and diligent practiser. Thou onely, knowest all thinges precisely (O God) who hast made weight and Balance, thy Iudgement: who hast created all thinges in *Number, Waight, and Measure*: and hast wayed the mountaines and hils in a Balance: who hast peysed in thy hand, both Heauen and earth. We therfore warned by the Sacred word, to Consider thy Creatures: and by that consideration, to wynne a glyms (as it were,) or shaddow of perceiuerance, that thy wisedome, might, and goodnes is infinite, and vnspeakable, in thy Creatures declared: And being farder aduertised, by thy mercifull goodnes, that, three principall wayes, were, of the, vsed in Creation of all thy Creatures, namely, *Number*, *Waight* and *Measure*, And for as much as, of *Number* and *Measure*, the two Artes (auncient, famous, and to humaine vses most necessary,) are, all ready, sufficiently knowen and extant: This third key, we beseche thee (through thy accustomed goodnes,) that it may come to the nedefull and sufficient knowledge, of such thy Seruauntes, as in thy workemanship, would gladly finde, thy true occasions (purposely of the vsed) whereby we should glorifie thy name, and shew forth (to the weaklinges in faith) thy wondrous wisedome and Goodnes. Amen.

Meruaile nothing at this pang (godly frend, you Gentle and zelous Student.) An other day, perchaunce, you will perceiue, what occasion moued me. Here, as now, I will giue you some ground, and withall some shew, of certaine commodities, by this Arte arising. And bycause this Arte is rare, my wordes and practises might be to darke: vnleast you had some light, holden before the matter: and that, best will be, in giuing you, out of *Archimedes* demonstrations, a few principal Conclusions, as foloweth.

1.

The Superficies of euery Liquor, by it selfe consistyng, and in quyet, is Sphæricall: the centre whereof, is the same, which is the centre of the Earth.

2.

If Solide Magnitudes, being of the same bignes, or quãtitie, that any Liquor is, and hauyng also the same Waight: be let downe into the same Liquor, they will settle downeward, so, that no parte of them, shall be aboue the Superficies of the Liquor: and yet neuertheles, they will not sinke vtterly downe, or drowne.

3.

If any Solide Magnitude beyng Lighter then a Liquor, be let downe into the same Liquor, it will settle downe, so farre into the same Liquor, that so great a quantitie of that Liquor, as is the parte of the Solid Magnitude, settled downe into the same Liquor: is in Waight, æquall, to the waight of the whole Solid Magnitude.

4.

Any Solide Magnitude, Lighter then a Liquor, forced downe into

into the ſame Liquor, will moue vpward, with ſo great a power, by how much, the Liquor hauyng æquall quantitie to the whole Magnitude, is heauyer then the ſame Magnitude.

5.

Any Solid Magnitude, heauyer then a Liquor, beyng let downe into the ſame Liquor, will ſinke downe vtterly: And wilbe in that Liquor, Lighter by ſo much, as is the waight or heauynes of the Liquor, hauing bygnes or quantitie, æquall to the Solid Magnitude.

6.

If any Solide Magnitude, Lighter then a Liquor, be let downe into the ſame Liquor, the waight of the ſame Magnitude, will be, to the Waight of the Liquor. (Which is æquall in quantitie to the whole Magnitude,) in that proportion, that the parte, of the Magnitude ſettled downe, is to the whole Magnitude.

I.D. The Cutting of a Sphere according to any proportion aſsigned, may by this propoſition be done Mechanically by tempering Liquor to a certayne waight in reſpect of the waight of the Sphare therein Swymming.

BY theſe verities, great Errors may be reformed, in Opinion of the Naturall Motion of thinges, Light and Heauy. Which errors, are in Naturall Philoſophie (almoſt) of all mẽ allowed: to much truſting to Authority: and falſe Suppoſitions. As, Of any two bodyes, the heauyer, to moue downward faſter then the lighter. This error, is not firſt by me, Noted: but by one *Iohn Baptiſt de Benedictis.* The chief of his propoſitions, is this: which ſeemeth a Paradox.

A common error noted.

If there be two bodyes of one forme, and of one kynde, æquall in quantitie or vnæquall, they will moue by æquall ſpace, in æquall tyme: So that both theyr mouynges be in ayre, or both in water: or in any one Middle.

A paradox.

Hereupon, in the feate of Gunnyng, certaine good diſcourſes (otherwiſe) may receiue great amendement, and furderance. In the entended purpoſe, alſo, allowing ſomwhat to the imperfection of Nature: not aunſwerable to the preciſenes of demonſtration. Moreouer, by the foreſaid propoſitions (wiſely vſed.) The Ayre, the water, the Earth, the Fire, may be nerely, knowen, how light or heauy they are (Naturally) in their aſsigned partes: or in the whole. And then, to thinges Elementall, turning your practiſe: you may deale for the proportion of the Elementes, in the thinges Compounded. Then, to the proportions of the Humours in Man: their waightes: and the waight of his bones, and fleſh. &c. Than, by waight, to haue conſideration of the Force of man, any maner of way: in whole or in part. Then, may you, of Ships water drawing, diuerſly, in the Sea and in freſh water, haue pleaſant conſideration: and of waying vp of any thing, ſonken in Sea or in freſh water &c. And (to lift vp your head a loft:) by waight, you may, as preciſely, as by any inſtrument els, meaſure the Diameters of *Sonne* and *Mone.&c.* Frende, I pray you, way theſe thinges, with the iuſt Balance of Reaſon. And you will finde Meruailes vpon Meruailes: And eſteme one Drop of Truth (yea in Naturall Philoſophie) more worth, then whole Libraries of Opinions, vndemonſtrated: or not aunſwering to Natures Law, and your experience. Leauing theſe

N. T. The wonderfull vſe of theſe Propoſitions.

 thinges,

thinges,thus: I will giue you two or three,light practises, to great purpose: and so finish my Annotation *Staticall.* In Mathematicall matters, by the Mechaniciens ayde, we will behold, here, the Commodity of waight. Make a Cube,of any one Vniforme: and through like heauy stuffe: of the same Stuffe,make a Sphære or Globe,precisely,of a Diameter æquall to the Radicall side of the Cube. Your stuffe,may be wood, Copper, Tinne, Lead,Siluer.&c. (being,as I sayd,of like nature, condition,and like waight throughout.) And you may, by Say Balance, haue prepared a great number of the smallest waightes: which, by those Balance can be discerned or tryed: and so,haue proceded to make you a perfect Pyle, company & Number of waightes: to the waight of six,eight,or twelue pound waight: most diligently tryed,all. And of euery one, the Content knowen, in your least waight,that is wayable. [They that can not haue these waightes of precisenes: may, by Sand,Vniforme,and well dusted,make them a number of waightes,somewhat nere precisenes: by halfing euer the Sand: they shall, at length, come to a least common waight. Therein,I leaue the farder matter,to their discretion,whom nede shall pinche.] The *Venetians* consideration of waight, may seme precise enough: by eight descentes progressionall,* halfing, from a grayne. Your Cube, Sphære,apt Balance,and conuenient waightes,being ready: fall to worke. ⁂. First, way your Cube. Note the Number of the waight. Way,after that, your Sphære. Note likewise,the Nūber of the waight. If you now find the waight of your Cube, to be to the waight of the Sphære,as 21. is to 11: Then you see, how the Mechanicien and *Experimenter*, without Geometrie and Demonstration, are (as nerely in effect) tought the proportion of the Cube to the Sphere: as I haue demonstrated it,in the end of the twelfth boke of *Euclide.* Often, try with the same Cube and Sphære. Then,chaunge,your Sphære and Cube,to an other matter: or to an other bignes: till you haue made a perfect vniuersall Experience of it. Possible it is, that you shall wynne to nerer termes,in the proportion.

The practise Staticall, to know the proportion, betwene the Cube, and the Sphære.

I.D.

**For, so, haue you. 256. partes of a Graine.*

When you haue found this one certaine Drop of Naturall veritie,procede on, to Inferre,and duely to make assay,of matter depending. As, bycause it is well demonstrated, that a Cylinder, whose heith, and Diameter of his base,is æquall to the Diameter of the Sphære, is Sesquialter to the same Sphære (that is,as 3. to 2:) To the number of the waight of the Sphære,adde halfe so much,as it is: and so haue you the number of the waight of that Cylinder. Which is also Comprehended of our former Cube: So,that the base of that Cylinder, is a Circle described in the Square, which is the base of our Cube. But the Cube and the Cylinder,being both of one heith, haue their Bases in the same proportion, in the which,they are, one to an other, in their Masines or Soliditie. But,before,we haue two numbers,expresing their Masines, Solidities, and Quantities, by waight: wherfore,we haue * the proportion of the Square,to the Circle, inscribed in the same Square. And so are we fallen into the knowledge sensible, and Experimentall of *Archimedes* great Secret: of him, by great trauaile of minde, sought and found. Wherfore,to any Circle giuen, you can giue a Square æquall: * as I haue taught,in my Annotation,vpon the first proposition of the twelfth boke, And likewise,to any Square giuen,you may giue a Circle æquall: *If you describe a Circle,which shall be in that proportion, to your Circle inscribed, as the Square is to the same Circle. This,you may do,by my Annotations,vpon the second proposition of the twelfth boke of *Euclide*, in my third Probleme there. Your diligence may come to a proportion,of the Square to the Circle inscribed, nerer the truth,then is the proportion of 14. to 11. And consider, that you may begyn at the Circle and Square, and so come to conclude of the Sphære,& the Cube,what

**The proportion of the Square to the Circle inscribed.*

**The Squaring of the Circle, Mechanically.*

**To any Square geuen, to geue a Circle, equall.*

their

their proportion is:as now, you came from the Sphære to the Circle. For,of Siluer,or Gold,or Latton Lamyns or plates(thorough one hole drawẽ,as the maner is)if you make a Square figure & way it:and then,describing theron, the Circle inscribed: & cut of,& file away,precisely (to the Circle) the ouerplus of the Square: you shall then,waying your Circle, see,whether the waight of the Square, be to your Circle, as 14. to 11. As I haue Noted, in the beginning of *Euclides* twelfth boke.&c.after this resort to my last proposition,vpon the last of the twelfth. And there,helpe your selfe,to the end. And, here, Note this, by the way. That we may Square the Circle, without hauing knowledge of the proportion,of the Circumference to the Diameter: as you haue here perceiued. And otherwayes also, I can demonstrate it.So that,many haue cumberd them selues superfluously, by trauailing in that point first, which was not of necessitie,first: and also very intricate. And easily,you may, (and that diuersly) come to the knowledge of the Circumference:the Circles Quantitie, being first knowen. Which thing,I leaue to your consideration:making hast to despatch an other Magistrall Probleme: and to bring it,nerer to your knowledge,and readier dealing with,then the world(before this day,)had it for you,that I can tell of.And that is, *A Mechanicall Dubblyng of the Cube:&c.* Which may, thus,be done: Make of Copper plates,or Tyn plates,a foursquare vpright Pyramis,or a Cone: perfectly fashioned in the holow,within. Wherin, let great diligence be vsed, to approche (as nere as may be) to the Mathematicall perfection of those figures. At their bases,let them be all open:euery where, els, most close,and iust to. From the vertex, to the Circumference of the base of the Cone: & to the sides of the base of the Pyramis:Let 4.straight lines be drawen,in the inside of the Cone and Pyramis: makyng at their fall,on the perimeters of the bases, equall angles on both sides them selues, with the sayd perimeters. These 4.lines (in the Pyramis:and as many,in the Cone)diuide:one,in 12. æquall partes: and an other, in 24.an other,in 60, and an other, in 100. (reckenyng vp from the vertex.) Or vse other numbers of diuision, as experience shall teach you. Then,* set your Cone or Pyramis, with the vertex downward, perpendicularly, in respect of the Base.(Though it be otherwayes,it hindreth nothyng.) So let thẽ most stedily be stayed. Now,if there be a Cube,which you wold haue Dubbled.Make you a prety Cube of Copper, Siluer, Lead, Tynne, Wood, Stone, or Bone. Or els make a hollow Cube,or Cubik coffen, of Copper,Siluer, Tynne,or Wood &c. These,you may so proportiõ in respect of your Pyramis or Cone, that the Pyramis or Cone, will be hable to conteine the waight of them, in water, 3. or 4. times:at the least: what stuff so euer they be made of.Let not your Solid angle, at the vertex,be to sharpe: but that the water may come with ease,to the very vertex,of your hollow Cone or Pyramis.Put one of your Solid Cubes in a Balance apt: take the waight therof exactly in water. Powre that water, (without losse) into the hollow Pyramis or Cone,quietly. Marke in your lines,what numbers the water Cutteth: Take the waight of the same Cube againe: in the same kinde of water, which you had before: put that* also, into the Pyramis or Cone,where you did put the first. Marke now againe, in what number or place of the lines, the water Cutteth them. Two

Note Squaring of the Circle without knowledge of the proportion betwene Circumference and Di[illegible]ter.

To Dubble the Cube redily: by Art Mechanicall: depending vppon Demonstration Mathematicall.

I. D. The 4.sides of this Pyramis must be 4. Isosceles Triangles a like and æquall.

*I. D. *In all workinges with this Pyramis or Cone, Let their Situations be in all Pointes and Conditions, a like, or all one: while you are about one worke. Els you will erre.*

*I. D. * Consider well when you must put your waters togyther: and when, you must empty your first water, out of your Pyramis or Cone. Els you will erre.*

wayes

wayes you may conclude your purpose : it is to wete, either by numbers or lines. By numbers : as, if you diuide the side of your Fundamentall Cube into so many æquall partes, as it is capable of, conueniently, with your ease, and precisenes of the diuision. For, as the number of your first and lesse line (in your hollow Pyramis or Cone,) is to the second or greater (both being counted from the vertex) so shall the number of the side of your Fundamentall Cube, be to the nūber belonging to the Radicall side, of the Cube, dubble to your Fundamentall Cube: Which being multiplied Cubik wise, will sone shew it selfe, whether it be dubble or no, to the Cubik number of your Fundamentall Cube. By lines, thus: As your lesse and first line, (in your hollow Pyramis or Cone,) is to the second or greater, so let the Radical side of your Fundamētall Cube, be to a fourth proportionall line, by the 12. proposition, of the sixth boke of *Euclide*. Which fourth line, shall be the Rote Cubik, or Radicall side of the Cube, dubble to your Fundamentall Cube : which is the thing we desired. For this, may I (with ioy) say, ΕΥΡΗΚΑ, ΕΥΡΗΚΑ, ΕΥΡΗΚΑ: thanking the holy and glorious Trinity: hauing greater cause therto, then * *Archimedes* had (for finding the fraude vsed in the Kinges Crowne, of Gold): as all men may easily Iudge : by the diuersitie of the frute following of the one, and the other. Where I spake before, of a hollow Cubik Coffen: the like vse, is of it: and without waight. Thus. Fill it with water, precisely full, and poure that water into your Pyramis or Cone. And here note the lines cutting in your Pyramis or Cone. Againe, fill your coffen, like as you did before. Put that Water, also, to the first. Marke the second cutting of your lines. Now, as you proceded before, so must you here procede. * And if the Cube, which you should Double, be neuer so great: you haue, thus, the proportion (in small) betwene your two litle Cubes: And then, the side, of that great Cube (to be doubled) being the third, will haue the fourth, found, to it proportionall: by the 12. of the sixth of Euclide.

** Vitruuius. Lib. 9. Cap. 3.* ☞ *God be thanked for this Inuention, & the fruite ensuing.* **Note.*

Note, that all this while, I forget not my first Proposition Staticall, here rehearsed: that, the Superficies of the water, is Sphæricall. Wherein, vse your discretion: to the first line, adding a small heare breadth, more: and to the second, halfe a heare breadth more, to his length. For, you will easily perceaue, that the difference can be no greater, in any Pyramis or Cone, of you to be handled. Which you shall thus trye. *For finding the swelling of the water aboue leuell.* Square the Semidiameter, from the Centre of the earth, to your first Waters Superficies. Square then, halfe the Subtendent of that watry Superficies (which Subtendent must haue the equall partes of his measure, all one, with those of the Semidiameter of the earth to your watry Superficies) : Subtracte this square, from the first : Of the residue, take the Rote Square. That Rote, Subtracte from your first Semidiameter of the earth to your watry Superficies : that, which remaineth, is the heith of the water, in the middle, aboue the leuell. Which, you will finde, to be a thing insensible. And though it were greatly sensible,* yet, by helpe of my sixt Theoreme vpon the last Proposition of Euclides twelfth booke, noted : you may reduce all, to a true Leuell. But, farther diligence, of you is to be vsed, against accidentall causes of the waters swelling: as by hauing (somwhat) with a moyst Sponge, before, made moyst your hollow Pyramis or Cone, will preuent an accidentall cause of Swelling, &c. Experience will teach you abundantly : with great ease, pleasure, and cōmoditie.

Note, as concerning the Sphæricall Superficies of the Water. ☞ **Note.*

Thus, may you Double the Cube Mechanically, Treble it, and so forth, in any proportion. Now will I Abridge your paine, cost, and Care herein. Without all preparing of your Fundamentall Cubes : you may (alike) worke this Conclusion. For, that, was rather a kinde of Experimentall demōstration, then the shortest way: and

Note this Abridgement of Dubbling the Cube. &c.

and all, vpon one Mathematicall Demonstration depending. Take water (as »
much as conueniently will serue your turne: as I warned before of your Funda- »
mentall Cubes bignes) Way it precisely. Put that water, into your Pyramis or »
Cone. Of the same kinde of water, then take againe, the same waight you had »
before: put that likewise into the Pyramis or Cone. For, in eche time, your mar- »
king of the lines, how the Water doth cut them, shall geue you the proportion be- »
twen the Radicall sides, of any two Cubes, wherof the one is Double to the other: »
working as before I haue taught you: *sauing that for you Fundamentall Cube his
Radicall side: here, you may take a right line, at pleasure.

Note. ☞

Yet farther proceding with our droppe of Naturall truth: you may (now) geue Cubes, one to the other, in any proportiō geuē: Rationall or Irrationall: on this maner. Make a hollow Parallelipipedon of Copper or Tinne: with one Base wāting, or open: as in our Cubike Coffen. Frō the bottome of that Parallelipipedon, raise vp, many perpendiculars, in euery of his fower sides. Now if

To giue Cubes one to the other in any proportion, Rationall or Irrationall.

any proportion be assigned you, in right lines: Cut one of your perpendiculars (or »
a line equall to it, or lesse then it) likewise: by the 10. of the sixth of Euclide. And »
those two partes, set in two sundry lines of those perpendiculars (or you may set »
them both, in one line) making their beginninges, to be, at the base: and so their »
lengthes to extend vpward. Now, set your hollow Parallelipipedon, vpright, »
perpendicularly, steadie. Poure in water, handsomly, to the heith of your shorter »
line. Poure that water, into the hollow Pyramis or Cone. Marke the place of »
the rising. Settle your hollow Parallelipipedon againe. Poure water into it: »
vnto the heith of the second line, exactly. Poure that water * duely into the »
hollow Pyramis or Cone: Marke now againe, where the water cutteth the same »
line which you marked before. For, there, as the first marked line, is to the se- »
cond: So shall the two Radicall sides be, one to the other, of any two Cubes: »
which, in their Soliditie, shall haue the same proportion, which was at the first as- »
signed: were it Rationall or Irrationall.

** Emptying the first.*

Thus, in sundry waies you may furnishe your selfe with such straunge and profitable matter: which, long hath bene wished for. And though it be Naturally done and Mechanically: yet hath it a good Demonstration Mathematicall. Which is this: Alwaies, you haue two Like Pyramids: or two Like Cones, in the proportions assigned: and like Pyramids or Cones, are in proportion, one to the other, in the proportion of their Homologall sides (or lines) tripled. Wherefore, if to the first, and second lines, found in your hollow Pyramis or Cone, you ioyne a third and a fourth, in continuall proportion: that fourth line, shall be to the first, as the greater Pyramis or Cone, is to the lesse: by the 33. of the eleuenth of Euclide. If Pyramis to Pyramis, or Cone to Cone, be double, then shall * Line to Line, be also double, &c. But, as our first line, is to the second, so is the Radicall side of our Fundamentall Cube, to the Radicall side of the Cube to be made, or to be doubled: and therefore, to those twaine also, a third and a fourth line, in continuall proportion, ioyned: will geue the fourth line in that proportion to the first, as our fourth Pyramidall, or Conike line, was to his first: but that was double, or treble, &c. as the Pyramids or Cones were, one to an other (as we haue proued) therfore, this fourth, shalbe also double or treble to the first, as the Pyramids or Cones were one to an other: But our made Cube, is described of the second in proportion, of the fower proportionall lines: therfore * as the fourth line, is to the first, so is that Cube, to the first Cube: and we haue proued the fourth line, to be to the first, as the Pyramis or Cone, is to the Pyramis or Cone: Wherefore the Cube is

The demonstrations of this Dubbling of the Cube, and of the rest.

I.D.

** Hereby, helpe your self to become a precise practiser. And so consider, how, nothing at all, you are hindred (sensibly) by the Conuexitie of the water.*

** By the 33. of the eleuenth booke of Euclide.*

 to the

I.D.
**And your diligence in practise,can so (in waight of water)performe it: Therefore,now,you are able to geue good reason of your whole doing.*

to the Cube,as Pyramis is to Pyramis,or Cone is to Cone. But we* Suppose Pyramis to Pyramis,or Cone to Cone, to be double or treble.&c. Therfore Cube,is to Cube,double,or treble,&c. Which was to be demonstrated. And of the Parallelipipedō,it is euidēt, that the water Solide Parallelipipedons,are one to the other, as their heithes are,seing they haue one base. Wherfore the Pyramids or Cones, made of those water Parallelipipedons,are one to the other,as the lines are(one to the other)betwene which,our proportion was assigned. But the Cubes made of lines,after the proportiō of the Pyramidal or Conik *homologall* lines,are one to the other,as the Pyramides or Cones are, one to the other (as we before did proue) therfore,the Cubes made, shalbe one to the other,as the lines assigned,are one to the other: Which was to be demonstrated. Note. *This,my Demonstratiō is more generall,then onely in Square Pyramis or Cone: Consider well. Thus, haue I, both Mathematically and Mechanically,ben very long in wordes: yet (I trust) nothing tedious to them,who, to these thinges, are well affected. And verily I am forced(auoiding prolixitie)to omit sundry such things,easie to be practised: which to the Mathematicien,would be a great Threasure: and to the Mechanicien,no small gaine.*Now may you, Betwene two lines giuen, finde two middle proportionals, in Continuall proportion: by the hollow Parallelipipedon, and the hollow Pyramis, or Cone. Now,any Parallelipipedon rectangle being giuen: thre right lines may be found,proportionall in any proportion assigned,of which,shal be produced a Parallelipipedon, æquall to the Parallelipipedon giuen. Hereof,I noted somwhat,vpon the 36,proposition,of the 11.boke of *Euclide*. Now,all those thinges,which *Vitruuius* in his Architecture,specified hable to be done, by dubbling of the Cube Or, by finding of two middle proportionall lines, betwene two lines giuen,may easely be performed. Now, that Probleme,which I noted vnto you,in the end of my Addition, vpon the 34.of the 11. boke of *Euclide*, is proued possible. Now,may any regular body,be Transformed into an other,&c. Now, any regular body: any Sphere, yea any Mixt Solid: and (that more is) Irregular Solides, may be made(in any proportiō assigned)like vnto the body,first giuen. Thus,of a *Manneken*,(as the *Dutch* Painters terme it)in the same *Symmetrie*, may a Giant be made: and that,with any gesture,by the Manneken vsed: and contrarywise. Now, may you, of any Mould, or Modell of a Ship, make one,of the same Mould (in any assigned proportion) bigger or lesser. Now, may you,of any*Gunne,or little peece of ordinaūce,make an other,with the same *Symmetrie* (in all pointes) as great,and as little,as you will. Marke that: and thinke on it. Infinitely, may you apply this, so long sought for,and now so easily concluded: and withall,so willingly and frankly communicated to such,as faithfully deale with vertuous studies. Thus,can the Mathematicall minde,deale Speculatiuely in his own Arte: and by good meanes, Mount aboue the cloudes and sterres: And thirdly,he can, by order,Descend,to frame Naturall thinges, to wonderfull vses: and when he list, retire home into his owne Centre: and there,prepare more Meanes,to Ascend or Descend by: and, all,to the glory of God, and our honest delectation in earth.

**Note this Corollary.*

**The great Commodities following of these new Inuentions.*

☞*

Such is the Fruite of the Mathematicall Sciences and Artes.

Although, the Printer, hath looked for this Præface,a day or two, yet could I not bring my pen from the paper, before I had giuen you comfortable warning, and brief instructions,of some of the Commodities,by *Statike*,hable to be reaped: In the rest,I will therfore,be as brief,as it is possible: and with all,describing them, somwhat accordingly. And that,you shall perceiue,by this, which in order commeth

meth next. For, wheras, it is so ample and wonderfull, that, an whole yeare long, one might finde fruitfull matter therin, to speake of: and also in practise, is a Threasure endeles: yet will I glanse ouer it, with wordes very few.

THis do I call **Anthropographie**. Which is an Art restored, and of my preferment to your Seruice. I pray you, thinke of it, as of one of the chief pointes, of Humane knowledge. Although it be, but now, first Cõfirmed, with this new name: yet the matter, hath from the beginning, ben in consideration of all perfect Philosophers. Anthropographie, is the description of the Number, Measure, Waight, figure, Situation, and colour of euery diuerse thing, conteyned in the perfect body of MAN: with certain knowledge of the Symmetrie, figure, waight, Characterization, and due locall motion, of any parcell of the sayd body, asigned: and of Nũbers, to the sayd parcell appertainyng. This, is the one part of the Definition, mete for this place: Sufficient to notifie, the particularitie, and excellency of the Arte: and why it is, here, ascribed to the Mathematicals. Yf the description of the heauenly part of the world, had a peculier Art, called *Astronomie*: If the description of the earthly Globe, hath his peculier arte, called *Geographie*. If the Matching of both, hath his peculier Arte, called *Cosmographie*: Which is the Descriptiõ of the whole, and vniuersall frame of the world: Why should not the description of him, who is the Lesse world: and, frõ the beginning, called *Microcosmus* (that is. *The Lesse World.*) And for whose sake, and seruice, all bodily creatures els, were created: Who, also, participateth with Spirites, and Angels: and is made to the Image and similitude of *God*: haue his peculier Art? and be called the *Arte of Artes*: rather, then, either to want a name, or to haue to base and impropre a name? You must of sundry professions, borow or challenge home, peculier partes hereof: and farder procede: as, God, Nature, Reason and Experience shall informe you. The Anatomistes will restore to you, some part: The Physiognomistes, some: The Chyromantistes some. The Metaposcopistes, some: The excellent, *Albert Durer*, a good part: the Arte of Perspectiue, will somwhat, for the Eye, helpe forward: *Pythagoras, Hipocrates, Plato, Galenus, Meletius*, & many other (in certaine thinges) will be Contributaries. And farder, the Heauen, the Earth, and all other Creatures, will eche shew, and offer their Harmonious seruice, to fill vp, that, which wanteth hereof: and with your own Experience, concluding: you may Methodically register the whole, for the posteritie: Whereby, good profe will be had, of our Harmonious, and Microcosmicall constitution. The outward Image, and vew hereof: to the Art of *Zographie* and Painting, to Sculpture, and Architecture: (for Church, House, Fort, or Ship) is most necessary and profitable: for that, it is the chiefe base and foundation of them. Looke in * *Vitruuius*, whether I deale sincerely for your behoufe, or no. Looke in *Albertus Durerus, De Symmetria humani Corporis*. Looke in the 27. and 28. Chapters, of the second booke, *De occulta Philosophia*. Consider the *Arke* of *Noe*. And by that, wade farther. Remember the *Delphicall Oracle* *NOSCE TEIPSUM* (*Knowe thy selfe*) so long agoe pronounced: of so many a Philosopher repeated: and of the *Wisest* attempted: And then, you will perceaue, how long agoe, you haue bene called to the Schole, where this Arte might be learned. Well. I am nothing affrayde, of the disdayne of some such, as thinke Sciences and Artes, to be but Seuen. Perhaps, those Such, may, with ignorance, and shame enough, come short of them Seuen also: and yet neuerthelesse

MAN is the Lesse World.

☿ Micro Cosmus.

**Lib. 3. Cap. 1.*

 they

they can not preſcribe a certaine number of Artes: and in eche, certaine vnpaſſable boundes, to God, Nature, and mans Induſtrie. New Artes, dayly riſe vp: and there was no ſuch order taken, that, All Artes, ſhould in one age, or in one land, or of one man, be made knowen to the world. Let vs embrace the giftes of God, and wayes to wiſedome, in this time of grace, from aboue, continually beſtowed on them, who thankefully will receiue them: *Et bonis Omnia Cooperabuntur in bonum.*

Trochilike, is that Art Mathematicall, which demonſtrateth the properties of all Circular motions, Simple and Compounde. And bycauſe the frute hereof, vulgarly receiued, is in Wheles, it hath the name of *Trochilike*: as a man would ſay, *Whele Art*. By this art, a Whele may be geuen which ſhall moue ones about, in any tyme aſſigned. Two Wheles may be giuen, whoſe turnynges about in one and the ſame tyme, (or equall tymes), ſhall haue, one to the other, any proportion appointed. By Wheles, may a ſtraight line be deſcribed: Likewiſe, a Spirall line in plaine, Conicall Section lines, and other Irregular lines, at pleaſure, may be drawen. Theſe, and ſuch like, are principall Concluſions of this Arte: and helpe forward many pleaſant and profitable Mechanicall workes: As Milles, to Saw great and very long Deale bordes, no man being by. Such haue I ſeene in Germany: and in the Citie of Prage: in the kingdome of Bohemia: Coyning Milles, Hand Milles for Corne grinding: And all maner of Milles, and Whele worke: By Winde, Smoke, Water, Waight, Spring, Man or Beaſt, moued. Take in your hand, *Agricola De re Metallica*: and then ſhall you (in all Mines) perceaue, how great nede is, of Whele worke. By Wheles, ſtraunge workes and incredible, are done: as will, in other Artes hereafter, appeare. A wonderfull example of farther poſſibilitie, and preſent commoditie, was ſene in my time, in a certaine Inſtrument: which by the Inuenter and Artificer (before) was ſolde for xx. Talentes of Golde: and then had (by miſfortune) receaued ſome iniurie and hurt: And one *Ianellus* of *Cremona* did mend the ſame, and preſented it vnto the Emperour *Charles* the fifth. *Hieronymus Cardanus*, can be my witneſſe, that therein, was one Whele, which moued, and that, in ſuch rate, that, in 7000. yeares onely, his owne periode ſhould be finiſhed. A thing almoſt incredible: But how farre, I keepe me within my boundes: very many men (yet aliue) can tell.

Saw Milles.

Helicoſophie, is nere Siſter to *Trochilike*: and is, An Arte Mathematicall, which demonſtrateth the deſigning of all Spirall lines in Plaine, on Cylinder, Cone, Sphære, Conoid, and Sphæroid, and their properties appertayning. The vſe hereof, in *Architecture*, and diuerſe Inſtrumentes and Engines, is moſt neceſſary. For, in many thinges, the Skrue worketh the feate, which, els, could not be performed. By helpe hereof, it is * recorded, that, where all the power of the Citie of Syracuſa, was not hable to moue a certaine Ship (being on ground) mightie *Archimedes*, ſetting to, his Skruiſh Engine, cauſed *Hiero* the king, by him ſelf, at eaſe, to remoue her, as he would. Wherat, the King wondring: Ἀπὸ ταύτης τῆς ἡμέρας, περὶ παντὸς, Ἀρχιμήδει λέγοντι πιστευτέον. *From this day, forward* (ſaid the King) *Credit ought to be giuen to Archimedes, what ſoeuer he ſayth.*

** Atheneus Lib.5.cap.8.*

Proclus. Pag.18.

Pneumatithmie demonſtrateth by cloſe hollow Geometricall Figures, (regular and irregular) the ſtraunge properties (in motion or ſtay) of the Water, Ayre, Smoke, and Fire, in theyr cōtinuitie, and

and as they are ioyned to the Elementes next them. This Arte, to the Naturall Philosopher, is very proffitable: to proue, that *Vacuum*, or *Emptines* is not in the world. And that, all Nature, abhorreth it so much: that, contrary to ordinary law, the Elementes will moue or stand: As, Water to ascend: rather then betwene him and Ayre, Space or place should be left, more then (naturally) that quãtitie of Ayre requireth, or can fill. Againe, Water to hang, and not descend: rather then by descending, to leaue Emptines at his backe. The like, is of Fire and Ayre: they will descend: when, either, their Cõtinuitie should be dissolued: or their next Element forced from them. And as they will not be extended, to discontinuitie: So, will they not, nor yet of mans force, can be prest or pent, in space, not sufficient and aunswerable to their bodily substance. Great force and violence will they vse, to enioy their naturall right and libertie. Hereupon, two or three men together, by keping Ayre vnder a great Cauldron, and forcyng the same downe, orderly, may without harme descend to the Sea bottome: and continue there a tyme &c. Where, Note, how the thicker Element (as the Water) giueth place to the thynner (as, is the ayre:) and receiueth violence of the thinner, in maner. &c. Pumps and all maner of Bellowes, haue their ground of this Art: and many other straunge deuises. As, *Hydraulica*, Organes goyng by water. &c. Of this Feat, (called commonly *Pneumatica*,) goodly workes are extant, both in Greke, and Latin. With old and learned Schole men, it is called *Scientia de pleno & vacuo.*

To go to the bottom of the Sea without daunger.

Menadrie, is an Arte Mathematicall, which demonstrateth, how, aboue Natures vertue and power simple: Vertue and force may be multiplied: and so, to direct, to lift, to pull to, and to put or cast fro, any multiplied or simple, determined Vertue, Waight or Force: naturally, not, so, directible or moueable. Very much is this Art furdred by other Artes: as, in some pointes, by *Perspectiue*: in some, by *Statike*: in some, by *Trochilike*: and in other, by *Helicosophie*: and *Pneumatithmie*. By this Art, all Cranes, Gybbettes, & Ingines to lift vp, or to force any thing, any maner way, are ordred: and the certaine cause of their force, is knowne: As, the force which one man hath with the Duche waghen Racke: therwith, to set vp agayne, a mighty waghen laden, being ouerthrowne. The force of the Crossebow Racke, is certainly, here, demonstrated. The reason, why one mã, doth with a leauer, lift that, which Sixe men, with their handes onely, could not, so easily do. By this Arte, in our common Cranes in London, where powre is to Crane vp, the waight of 2000. pound: by two Wheles more (by good order added) Arte concludeth, that there may be Craned vp 200000. pound waight &c. So well knew *Archimedes* this Arte: that he alone, with his deuises and engynes, (twise or thrise) spoyled and discomfited the whole Army and Hoste of the Romaines, besieging *Syracusa*, *Marcus Marcellus the Consul*, being their Generall Capitaine. Such huge Stones, so many, with such force, and so farre, did he with his engynes hayle among them, out of the Citie. And by Sea likewise: though their Ships might come to the walls of *Syracusa*, yet hee vtterly confounded the Romaine Nauye. What with his mighty Stones hurlyng: what with Pikes of* 18 fote long, made like shaftes: which he forced almost a quarter of a myle. What, with his catchyng hold of their Shyps, and hoysing them vp aboue the water, and suddenly letting them fall into the Sea againe: what with his* Burning Glasses: by which he fired their other Shippes a farof: what, with his other pollicies, deuises, and engines, he so manfully acquit him selfe: that all the Force, courage, and pollicie of the Romaines (for a great season)

Plutarchus in Marco Marcello.
Synesius in Epistolis.
Polybius.
Plinius.
Quintilianus.
T. Liuius.
** Athenæus.*

** Galenus.*
Anthemius.

 could

could nothing preuaile, for the winning of Syracusa. Wherupon, the Romanes named *Archimedes, Briareus*, and *Centimanus*. *Zonaras* maketh mention of one *Proclus*, who so well had perceiued *Archimedes* Arte of *Menadrie*, and had so well inuented of his owne, that with his Burning Glasses, being placed vpon the walles of Bysance, he multiplied so the heate of the Sunne, and directed the beames of the same against his enemies Nauie with such force, and so sodeinly (like lightening) that he burned and destroyed both man and ship. And *Dion* specifieth of *Priscus*, a *Geometricien* in Bysance, who inuented and vsed sondry Engins, of Force multiplied: Which was cause, that the *Emperour Seuerus* pardoned him, his life, after he had wonne Bysance: Bycause he honored the Arte, wytt, and rare industrie of *Priscus*. But nothing inferior to the inuention of these engines of Force, was the inuention of Gunnes. Which, from an English man, had the occasion and order of first inuenting: though in an other land, and by other men, it was first executed. And they that should see the record, where the occasion and order generall, of Gunning, is first discoursed of, would thinke: that, small thinges, slight, and comon: comming to wise mens consideration, and industrious mens handling, may grow to be of force incredible.

Burning Glasses.

Gunnes.

Hypogeiodie, is an Arte Mathematicall, demonstratyng, how, vnder the Sphæricall Superficies of the earth, at any depth, to any perpendicular line assigned (whose distance from the perpendicular of the entrance: and the Azimuth, likewise, in respect of the said entrance, is knowen) certaine way may be præscribed and gone: And how, any way aboue the Superficies of the earth designed, may vnder earth, at any depth limited, be kept: goyng alwayes, perpendicularly, vnder the way, on earth designed: And, contrarywise, Any way, (straight or croked,) vnder the earth, beyng giuen: vppon the vtface, or Superficies of the earth, to Lyne out the same: So, as, from the Centre of the earth, perpendiculars drawen to the Sphæricall Superficies of the earth, shall precisely fall in the Correspondent pointes of those two wayes. This, with all other Cases and circumstances herein, and appertenances, this Arte demonstrateth. This Arte, is very ample in varietie of Conclusions: and very profitable sundry wayes to the Common Wealth. The occasion of my Inuenting this Arte, was at the request of two Gentlemen, who had a certaine worke (of gaine) vnder ground: and their groundes did ioyne ouer the worke: and by reason of the crokednes, diuers depthes, and heithes of the way vnder ground, they were in doubt, and at controuersie, vnder whose ground, as then, the worke was. The name onely (before this) was of me published, *De Itinere Subterraneo*: The rest, be at Gods will. For Pioners, Miners, Diggers for Mettalls, Stone, Cole, and for secrete passages vnder ground, betwene place and place (as this land hath diuerse) and for other purposes, any man may easily perceaue, both the great fruite of this Arte, and also in this Arte, the great aide of Geometrie.

Hydragogie, demonstrateth the possible leading of Water, by Natures lawe, and by artificiall helpe, from any head (being a Spring, standing, or running Water) to any other place assigned.

Long

Long,hath this Arte bene in vse : and much thereof written : and very marueilous workes therein,performed : as may yet appeare,in Italy: by the Ruynes remaining of the Aqueductes . In other places,of Riuers leading through the Maine land, Nauigable many a Mile. And in other places,of the marueilous forcinges of Water to Ascend . which all,declare the great skill,to be required of him,who should in this Arte be perfecte, for all occasions of waters possible leading . To speake of the allowance of the Fall,for euery hundred foote: or of the Ventills(if the waters labour be farre,and great) I neede not : Seing, at hand(about vs)many expert men can sufficiently testifie, in effecte, the order : though the Demonstration of the Necessitie thereof, they know not : Nor yet, if they should be led, vp and downe, and about Mountaines, from the head of the Spring:and then,a place being assigned : and of them, to be demaunded, how low or high,that last place is, in respecte of the head, from which (so crokedly, and vp and downe) they be come: Perhaps,they would not, or could not, very redily,or nerely assoyle that question. *Geometrie* therefore, is necessary to *Hydragogie* . Of the sundry wayes to force water to ascend , eyther by *Tympane, Kettell mills, Skrue, Ctesibike,* or such like : in *Vitruuius, Agricola,* (and other,)fully,the maner may appeare . And so,thereby,also be most euident, how the Artes, of *Pneumatithmie,Helicosophie, Statike , Trochilike,* and *Menadrie,* come to the furniture of this,in Speculation, and to the Commoditie of the Common Wealth,in practise.

Horometrie, is an Arte Mathematicall, which demõstrateth, how,at all times appointed, the precise vsuall denominatiõ of time, may be knowen,for any place assigned . These wordes,are smoth and plaine easie Englishe, but the reach of their meaning,is farther, then you woulde lightly imagine . Some part of this Arte, was called in olde time,*Gnomonice:* and of late,*Horologiographia :* and in Englishe,may be termed,*Dialling* . Auncient is the vse , and more auncient,is the Inuention . The vse,doth well appeare to haue bene (at the least) aboue two thousand and three hundred yeare agoe : in * King *Achaz* Diall, then,by the Sunne,shewing the distinction of time . By Sunne, Mone,and Sterres,this Dialling may be performed,and the precise Time of day or night knowen . But the demonstratiue delineation of these Dialls,of all sortes, requireth good skill,both of *Astronomie*,and *Geometrie* Elementall,Sphæricall,Phænomenall,and Conikall . Then,to vse the groundes of the Arte, for any regular Superficies, in any place offred : and (in any possible apt position therof) theron, to describe (all maner of wayes) how, vsuall howers, may be (by the *Sunnes* shadow) truely determined : will be found no sleight Painters worke . So to Paint, and prescribe the Sunnes Motion,to the breadth of a heare. In this Feate(in my youth) I Inuented a way,How in any Horizontall,Murall,or Æquinoctiall Diall,&c. At all howers(the Sunne shining)the Signe and Degree ascendent,may be knowen . Which is a thing very necessary for the Rising of those fixed Sterres : whose Operation in the Ayre, is of great might, euidently . I speake no further,of the vse hereof. But forasmuch as,Mans affaires require knowledge of Times & Momentes,when,neither Sunne,Mone,or Sterre, can be sene: Therefore,by Industrie Mechanicall, was inuented,first,how,by Water,running orderly,the Time and howers might be knowen: whereof, the famous *Ctesibius,* was Inuentor : a man, of *Vitruuius,* to the Skie (iustly) extolled . Then, after that, by Sand running, were howers measured : Then, by *Trochilike* with waight : And of late time, by *Trochilike* with Spring : without waight. All these,

* *4.Reg.20.*

 by

by Sunne or Sterres direction (in certaine time) require ouersight and reformation, according to the heauenly Æquinoctiall Motion: besides the inæqualitie of their owne Operation. There remayneth (without parabolicall meaning herein) among the Philosophers, a more excellent, more commodious, and more marueilous way, then all these: of hauing the motion of the Primouant (or first æquinoctiall motion,) by Nature and Arte, Imitated: which you shall (by furder search in waightier studyes) hereafter, vnderstand more of. And so, it is tyme to finish this Annotation, of Tymes distinction, vsed in our common, and priuate affaires: The commoditie wherof, no man would want, that can tell, how to bestow his tyme.

A perpetuall Motion.

Zographie, is an Arte Mathematicall, which teacheth and demonstrateth, how, the Intersection of all visuall Pyramides, made by any playne assigned, (the Centre, distance, and lightes, beyng determined) may be, by lynes, and due propre colours, represented. A notable Arte, is this: and would require a whole Volume, to declare the property thereof: and the Commodities ensuyng. Great skill of *Geometrie*, *Arithmetike*, *Perspectiue*, and *Anthropographie*, with many other particular Artes, hath the *Zographer*, nede of, for his perfection. For, the most excellent Painter, (who is but the propre Mechanicien, & Imitator sensible, of the Zographer) hath atteined to such perfection, that Sense of Man and beast, haue iudged thinges painted, to be things naturall, and not artificiall: aliue, and not dead. This Mechanicall Zographer (commonly called the Painter) is meruailous in his skill: and seemeth to haue a certaine diuine power: As, of frendes absent, to make a frendly, present comfort: yea, and of frendes dead, to giue a continuall, silent presence: not onely with vs, but with our posteritie, for many Ages. And so procedyng, Consider, How, in Winter, he can shew you, the liuely vew of Sommers Ioy, and riches: and in Sommer, exhibite the countenance of Winters dolefull State, and nakednes. Cities, Townes, Fortes, Woodes, Armyes, yea whole Kingdomes (be they neuer so farre, or greate) can he, with ease, bring with him, home (to any mans Iudgement) as Paternes liuely, of the thinges rehearsed. In one little house, can he, enclose (with great pleasure of the beholders,) the portrayture liuely, of all visible Creatures, either on earth, or in the earth, liuing: or in the waters lying, Creping, slyding, or swimming: or of any foule, or fly, in the ayre flying. Nay, in respect of the Starres, the Skie, the Cloudes: yea, in the shew of the very light it selfe (that Diuine Creature) can he match our eyes Iudgement, most nerely. What a thing is this? thinges not yet being, he can represent so, as, at their being, the Picture shall seame (in maner) to haue Created them. To what Artificer, is not Picture, a great pleasure and Commoditie? Which of them all, will refuse the Direction and ayde of Picture? The Architect, the Goldsmith, and the Arras Weauer: of Picture, make great account. Our liuely Herbals, our portraitures of birdes, beastes, and fishes: and our curious Anatomies, which way, are they most perfectly made, or with most pleasure, of vs beholden? Is it not, by Picture onely? And if Picture, by the Industry of the Painter, be thus commodious and meruailous: what shall be thought of *Zographie*, the Scholemaster of Picture, and chief gouernor? Though I mencion not *Sculpture*, in my Table of Artes Mathematicall: yet may all men perceiue, How, that *Picture* and *Sculpture*, are Sisters germaine: and both, right profitable, in a Commō wealth. and of *Sculpture*, aswell as of Picture, excellent Artificers haue written great bokes in commendation. Witnesse I take, of *Georgio Vasari, Pittore Aretino*: of *Pomponius Gauricus*: and other. To these two Artes, (with other,) is a certaine od Arte, called *Althalmasat*, much beholdyng: more, then the common *Sculptor, Entayler, Keruer, Cutter, Grauer, Founder,*

der,

der, or *Paynter* (*&c*) know their Arte, to be commodious.

Architecture, to many may seme not worthy, or not mete, to be reckned among the *Artes Mathematicall*. To whom, I thinke good, to giue some account of my so doyng. Not worthy, (will they say,) bycause it is but for building, of a house, Pallace, Church, Forte, or such like, grosse workes. And you, also, defined the *Artes Mathematicall*, to be such, as dealed with no Materiall or corruptible thing: and also did demonstratiuely procede in their faculty, by Number or Magnitude. First, you see, that I count, here, *Architecture*, among those *Artes Mathematicall*, which are Deriued from the Principals: and you know, that such, may deale with Naturall thinges, and sensible matter. Of which, some draw nerer, to the Simple and absolute Mathematicall Speculation, then other do. And though, the *Architect* procureth, enformeth, & directeth, the *Mechanicien*, to handworke, & the building actuall, of house, Castell, or Pallace, and is chief Iudge of the same: yet, with him selfe (as chief *Master* and *Architect*,) remaineth the Demonstratiue reason and cause, of the Mechaniciens worke: in Lyne, plaine, and Solid: by *Geometricall*, *Arithmeticall*, *Opticall*, *Musicall*, *Astronomicall*, *Cosmographicall* (& to be brief) by all the former Deriued *Artes Mathematicall*, and other Naturall Artes, hable to be confirmed and stablished. If this be so: then, may you thinke, that *Architecture*, hath good and due allowance, in this honest Company of *Artes Mathematicall* Deriuatiue. I will, herein, craue Iudgement of two most perfect *Architectes*: the one, being *Vitruuius*, the Romaine: who did write ten bookes thereof, to the Emperour *Augustus* (in whose daies our Heauenly Archemaster, was borne): and the other, *Leo Baptista Albertus*, a Florentine: who also published ten bookes therof. *Architectura* (sayth *Vitruuius*) *est Scientia pluribus disciplinis & varijs eruditionibus ornata: cuius Iudicio probantur omnia, quæ ab cæteris Artificibus perficiuntur opera*. That is. Architecture, is a Science garnished with many doctrines & diuerse instructions: by whose Iudgement, all workes, by other workmen finished, are Iudged. It followeth. *Ea nascitur ex Fabrica, & Ratiocinatione. &c. Ratiocinatio autem est, quæ, res fabricatas, Solertia ac ratione proportionis, demonstrare atque explicare potest*. *Architecture, groweth of Framing, and Reasoning. &c. Reasoning, is that, which of thinges framed, with forecast, and proportion: can make demonstration, and manifest declaration*. Agaand. *Cùm, in omnibus enim rebus, tùm maximè etiam in Architectura, hæc duo insunt: quod significatur, & quod significat. Significatur proposita res, de qua dicitur: hanc autem Significat Demonstratio, rationibus doctrinarum explicata*. *Forasmuch as, in all thinges: therefore chiefly in Architecture, these two thinges are: the thing signified: and that which signifieth. The thing propounded, whereof we speake, is the thing Signified. But Demonstration, expressed with the reasons of diuerse doctrines, doth signifie the same thing*. After that. *Vt literatus sit, peritus Graphidos, eruditus Geometriæ, & Optices non ignarus: instructus Arithmetica: historias complures nouerit, Philosophos diligenter audiuerit: Musicam sciuerit: Medicinæ non sit ignarus, responsa Iurisperitorũ nouerit: Astrologiam, Cælique rationes cognitas habeat*. *An Architect* (sayth he) *ought to vnderstand Languages, to be skilfull of Painting, well instructed in Geometrie, not ignorant of Perspectiue, furnished with Arithmetike, haue knowledge of many histories, and diligently haue heard Philosophers, haue skill of Musike, not ignorant of Physike, know the aunsweres of Lawyers, and haue Astro-*

An obiection.

The Answer.

 nomie,

nomie, and the courses Cælestiall, in good knowledge. He geueth reason, orderly, wherefore all these Artes, Doctrines, and Instructions, are requisite in an excellent *Architect*. And (for breuitie) omitting the Latin text, thus he hath. *Secondly, it is behofefull for an Architect to haue the knowledge of Painting: that he may the more easilie fashion out, in patternes painted, the forme of what worke he liketh. And Geometrie, geueth to Architecture many helpes: and first teacheth the Vse of the Rule, and the Cumpasse: wherby (chiefly and easilie) the descriptions of Buildinges, are despatched in Groundplats: and the directions of Squires, Leuells, and Lines. Likewise, by Perspectiue, the Lightes of the heauen, are well led, in the buildinges: from certaine quarters of the world. By Arithmetike, the charges of Buildinges are summed together: the measures are expressed, and the hard questions of Symmetries, are by Geometricall Meanes and Methods discoursed on. &c. Besides this, of the Nature of thinges (which in Greke is called* φυσιολογία*) Philosophie doth make declaration. Which, it is necessary, for an Architect, with diligence to haue learned: because it hath many and diuers naturall questions: as specially, in Aqueductes. For in their courses, leadinges about, in the leuell ground, and in the mountinges, the naturall Spirites or breathes are ingendred diuers wayes: The hindrances, which they cause, no man can helpe, but he, which out of Philosophie, hath learned the originall causes of thinges. Likewise, whosoeuer shall read Ctesibius, or Archimedes bookes, (and of others, who haue written such Rules) can not thinke, as they do: vnlesse he shall haue receaued of Philosophers, instructions in these thinges. And Musike he must nedes know: that he may haue vnderstanding, both of Regular and Mathematicall Musike: that he may temper well his Balistes, Catapultes, and Scorpions. &c. Moreouer, the Brasen Vessels, which in Theatres, are placed by Mathematicall order, in ambries, vnder the steppes: and the diuersities of the soundes (which ŷ Grecians call* ἠχεῖα*) are ordred according to Musicall Symphonies & Harmonies: being distributed in ŷ Circuites, by Diatessaron, Diapente, and Diapason. That the conuenient voyce, of the players sound, whẽ it came to these preparations, made in order, there being increased: with ŷ increasing, might come more cleare & pleasant, to ŷ eares of the lokers on. &c. And of Astronomie, is knowẽ ŷ East, West, South, and North. The fashion of the heauen, the Æquinox, the Solsticie, and the course of the sterres. Which thinges, vnleast one know: he can not perceiue, any thyng at all, the reason of Horologies. Seyng therfore this ample Science, is garnished, beautified and stored, with so many and sundry skils and knowledges: I thinke, that none can iustly account them selues Architectes, of the suddeyne. But they onely, who from their childes yeares, ascendyng by these degrees of knowledges, beyng fostered vp with the atteynyng of many Languages and Artes, haue wonne to the high Tabernacle of Archicture. &c. And to whom Nature hath giuen such quicke Circumspection, sharpnes of witt, and Memorie, that they may be very absolutely skillfull in Geometrie, Astronomie, Musike, and the rest of the Artes Mathematicall:*

call: Such, surmount and passe the callyng, and state, of Architectes: and are become Mathematiciens. &c. And they are found, seldome. As, in tymes past, was Aristarchus Samius: Philolaus, and Archytas, Tarentynes: Apollonius Pergeus: Eratosthenes Cyreneus: Archimedes, and Scopas, Syracusians. Who also, left to theyr posteritie, many Engines and Gnomonicall workes: by numbers and naturall meanes, inuented and declared.

A Mathematicien.

Thus much, and the same wordes (in sense) in one onely Chapter of this Incõparable *Architect Vitruuius*, shall you finde. And if you should, but take his boke in your hand, and slightly loke thorough it, you would say straight way: This is *Geometrie, Arithmetike, Astronomie, Musike, Anthropographie, Hydragogie, Horometrie. &c.* and (to cõclude) the Storehouse of all workmãship. Now, let vs listen to our other Iudge, our Florentine, *Leo Baptista:* and narrowly consider, how he doth determine of *Architecture. Sed anteq̃ vltra progrediar. &c. But before I procede any further* (sayth he) *I thinke, that I ought to expresse, what man I would haue to bee allowed an Architect. For, I will not bryng in place a Carpenter: as though you might Compare him to the Chief Masters of other Artes. For the hand of the Carpenter, is the Architectes Instrument. But I will appoint the Architect to be that man, who hath the skill, (by a certaine and merueilous meanes and way,) both in minde and Imagination to determine: and also in worke to finish: what workes so euer, by motion of waight, and cuppling and framyng together of bodyes, may most aptly be Commodious for the worthiest Vses of Man. And that he may be able to performe these thinges, he hath nede of atteynyng and knowledge of the best, and most worthy thynges. &c. The whole Feate of Architecture in buildyng, consisteth in Lineamentes, and in Framyng. And the whole power and skill of Lineamentes, tendeth to this: that the right and absolute way may be had, of Coaptyng and ioyning Lines and angles: by which, the face of the buildyng or frame, may be comprehended and concluded. And it is the property of Lineamentes, to prescribe vnto buildynges, and euery part of them, an apt place, & certaine nũber: a worthy maner, and a semely order: that, so, ẏ whole forme and figure of the buildyng, may rest in the very Lineamentes. &c. And we may prescribe in mynde and imagination the whole formes,* * *all materiall stuffe beyng secluded. Which point we shall atteyne, by Notyng and forepointyng the angles, and lines, by a sure and certaine direction and connexion. Seyng then, these thinges, are thus: Lineamente, shalbe the certaine and constant prescribyng, conceiued in mynde: made in lines and angles: and finished with a learned minde and wyt.* We thanke you Master *Baptist*, that you haue so aptly brought your Arte, and phrase therof, to haue some Mathematicall perfection: by certaine order, nũber, forme, figure, and *Symmetrie* mentall: all naturall & sensible stuffe set a part. Now, then, it is euident, (Gentle reader) how aptely and worthely, I haue preferred *Architecture*, to be bred and fostered vp in the Dominion of the pereles Princesse, *Mathematica:* and to be a naturall Subiect of hers. And the name of *Architecture*, is of the principalitie, which this Science hath, aboue all other Artes. And *Plato* affirmeth, the *Architect* to be *Master* ouer all, that make any worke. Wherupon, he is neither Smith, nor Builder: nor, separately, any Artificer: but the

Vitruuius.

VVho is an Architect.

* *The Immaterialitie of perfect Architecture.*

What, Lineament is.

Note.

Hed,the Prouoſt, the Directer,and Iudge of all Artificiall workes, and all Artificers. For,the true *Architect*,is hable to teach,Demonſtrate,diſtribute,deſcribe, and Iudge all workes wrought. And he,onely,ſearcheth out the cauſes and reaſons of all Artificiall thynges. Thus excellent,is *Architecture*: though few (in our dayes) atteyne thereto: yet may not the Arte,be otherwiſe thought on, then in very dede it is worthy. Nor we may not,of auncient Artes,make new and imperfect Definitions in our dayes: for ſcarſitie of Artificers: No more,than we may pynche in,the Definitions of *Wiſedome*,or *Honeſtie*, or of *Frendeſhyp* or of *Iuſtice*. No more will I conſent,to Diminiſh any whit,of the perfection and dignitie, (by iuſt cauſe) allowed to abſolute *Architecture*. Vnder the Direction of this Arte, are thre principall,neceſſary *Mechanicall Artes*. Namely, *Howſing*, *Fortification*, and *Naupegie*. *Howſing*, I vnderſtand,both for Diuine Seruice,and Mans common vſage: publike, and priuate. Of *Fortification* and *Naupegie*, ſtraunge matter might be told you: But perchaunce,ſome will be tyred,with this Bederoll, all ready rehearſed: and other ſome, will nycely nip my groſſe and homely diſcourſing with you: made in poſt haſt: for feare you ſhould wante this true and frendly warnyng, and taſt giuyng, of the *Power Mathematicall*. Lyfe is ſhort, and vncertaine: Tymes are perilouſe: &c. And ſtill the Printer awayting, for my pen ſtaying: All theſe thinges,with farder matter of Ingratefulnes, giue me occaſion to paſſe away, to the other Artes remainyng, with all ſpede poſsible.

THe Arte of Nauigation, demonſtrateth how,by the ſhorteſt good way, by the apteſt Directiõ,& in the ſhorteſt time, a ſufficient Ship,betwene any two places (in paſſage Nauigable,) aſsigned: may be cõducted: and in all ſtormes,& naturall diſturbances chauncyng, how,to vſe the beſt poſsible meanes, whereby to recouer the place firſt aſsigned. What nede, the *Maſter Pilote*, hath of other Artes, here before recited,it is eaſie to know: as, of *Hydrographie*, *Aſtronomie*, *Aſtrologie*, and *Horometrie*. Preſuppoſing continually,the common Baſe,and foundacion of all: namely *Arithmetike* and *Geometrie*. So that,he be hable to vnderſtand,and Iudge his own neceſſary Inſtrumentes,and furniture Neceſſary: Whether they be perfectly made or no: and alſo can,(if nede be) make them, hym ſelfe. As Quadrantes, The Aſtronomers Ryng,The Aſtronomers ſtaffe,The Aſtrolabe vniuerſall. An Hydrographicall Globe. Charts Hydrographicall,true,(not with parallell Meridians). The Common Sea Compas: The Compas of variacion: The Proportionall,and Paradoxall Compaſſes (of me Inuented,for our two Moſcouy Maſter Pilotes, at the requeſt of the Company) Clockes with ſpryng: houre,halfe houre,and three houre Sandglaſſes: & ſundry other Inſtrumẽtes: And alſo, be hable,on Globe, or Playne to deſcribe the Paradoxall Compaſſe: and duely to vſe the ſame,to all maner of purpoſes, whereto it was inuented. And alſo, be hable to Calculate the Planetes places for all tymes.

Anno. 1559.

Moreouer,with Sonne Mone or Sterre (or without) be hable to define the Longitude & Latitude of the place,which he is in: So that,the Longitude & Latitude of the place,from which he ſayled,be giuen: or by him,be knowne. whereto,appertayneth expert meanes,to be certified euer,of the Ships way. &c. And by foreſeing the Riſing,Settyng, Noneſtedyng, or Midnightyng of certaine tempeſtuous fixed Sterres: or their Coniunctions, and Anglynges with the Planetes, &c. he ought to haue expert coniecture of Stormes, Tempeſtes, and Spoutes: and ſuch lyke Meteorologicall effectes,daungerous on Sea. For (as *Plato* ſayth,) *Mutationes*,

oppor-

opportunitatesq́, temporum presentire, non minus rei militari, quàm Agriculturæ, Nauigationiq́, conuenit. To foresee the alterations and opportunities of tymes, is conuenient, no lesse to the Art of Warre, then to Husbandry and Nauigation. And besides such cunnyng meanes, more euident tokens in Sonne and Mone, ought of hym to be knowen: such as (the Philosophicall Poëte) *Virgilius* teacheth, in hys *Georgikes.* Where he sayth,

Sol quoq́, & exoriens & quum se condet in vndas,
Signa dabit, Solem certißima signa sequuntur. &c.

——————— *Nam sæpe videmus,*
Ipsius in vultu varios errare colores.
Cæruleus, pluuiam denunciat, igneus Euros.
Sin maculæ incipient rutilo immiscerier igni,
Omnia tum pariter vento, nimbisq́, videbis
Feruere: non illa quisquam me nocte per altum
Ire, neq́, a terra moueat conuellere funem. &c.
Sol tibi signa dabit. Solem quis dicere falsum
Audeat? ——— *&c.*

Georgic. 1.

And so of Mone, Sterres, Water, Ayre, Fire, Wood, Stones, Birdes, and Beastes, and of many thynges els, a certaine Sympathicall forewarnyng may be had: sometymes to great pleasure and profit, both on Sea and Land. Sufficiently, for my present purpose, it doth appeare, by the premisses, how *Mathematicall*, the *Arte* of *Nauigation*, is: and how it nedeth and also vseth other *Mathematicall Artes*: And now, if I would go about to speake of the manifold Commodities, commyng to this Land, and others, by Shypps and *Nauigation*, you might thinke, that I catch at occasions, to vse many wordes, where no nede is.

Yet, this one thyng may I, (iustly) say. In *Nauigation*, none ought to haue greater care, to be skillfull, then our English Pylotes. And perchaunce, Some, would more attempt: And other Some, more willingly would be aydyng, if they wist certainely, What Priuiledge, God had endued this Iland with, by reason of Situation, most commodious for *Nauigation*, to Places most Famous & Riche. And though, (of* Late) a young Gentleman, a Courragious Capitaine, was in a great readynes, with good hope, and great causes of persuasion, to haue ventured, for a Discouerye, (either *Westerly*, by *Cape de Paramantia*: or *Esterly*, aboue *Noua Zemla*, and the *Cyremisses*) and was, at the very nere tyme of Attemptyng, called and employed otherwise (both then, and since,) in great good seruice to his Countrey, as the Irish Rebels haue* tasted: Yet, I say, (though the same Gentleman, doo not hereafter, deale therewith) Some one, or other, should listen to the Matter: and by good aduise, and discrete Circumspection, by little, and little, wynne to the sufficient knowledge of that Trade and Voyage: Which, now, I would be sory, (through Carelesnesse, want of Skill, and Courrage,) should remayne Vnknowne and vnheard of. Seyng, also, we are herein, halfe Challenged, by the learned, by halfe request, published. Therof, verely, might grow Commoditye, to this Land chiefly, and to the rest of the Christen Common wealth, farre passing all riches and worldly Threasure.

*Anno. 1567 S. H. G.

*Anno. 1569

Thaumaturgike, is that Art Mathematicall, which giueth certaine order to make straunge workes, of the sense to be perceiued, and of men greatly to be wondred at. By sundry meanes, this *Wonderworke* is wrought. Some, by *Pneumatithmie*. As the workes of *Ctesibius* and *Hero*,

Some by waight.wherof *Timæus* speaketh.Some,by Stringes strayned,or Springs, therwith Imitating liuely Motions.Some, by other meanes,as the Images of Mercurie:and the brasen hed,made by *Albertus Magnus*,which dyd seme to speake.*Boethius* was excellent in these feates. To whom,*Cassiodorus* writyng,sayth.*Your purpose is to know profound thynges:and to shew meruayles. By the disposition of your Arte, Metals do low: Diomedes of brasse, doth blow a Trumpet loude: a brasen Serpent hisseth:byrdes made, sing swetely. Small thynges we rehearse of you,who can Imitate the heauen.&c.* Of the straunge Selfmouyng, which, at Saint Denys, by Paris, * I saw, ones or twise (*Orontius* beyng then with me, in Company)it were to straunge to tell. But some haue written it.And yet,(I hope) it is there,of other to be sene.And by *Perspectiue* also straunge thinges,are done.As partly(before)I gaue you to vnderstand in *Perspectiue*.As, to see in the Ayre, a loft, the lyuely Image of an other man, either walkyng to and fro : or standyng still. Likewise, to come into an house, and there to see the liuely shew of Gold, Siluer or precious stones:and commyng to take them in your hand, to finde nought but Ayre.Hereby, haue some men (in all other matters counted wise) fouly ouershot thē selues:misdeaming of the meanes.Therfore sayd *ClaudiusCælestinus. Hodie magnæ literaturæ viros & magnæ reputationis videmus, opera quedam quasi miranda, supra Naturā putare: de quibus in Perspectiua doctus causam faciliter reddidisset.* That is.*Now a dayes,we see some men,yea of great learnyng and reputation,to Iudge certain workes as meruaylous,aboue the power of Nature: Of which workes,one that were skillfull in Perspectiue might easely haue giuen the Cause.* Of *Archimedes Sphære,Cicero* witnesseth.Which is very straunge to thinke on. *For when Archimedes* (sayth he)*did fasten in a Sphære,the mouynges of the Sonne,Mone,and of the fiue other Planets,he did,as the God,which(in Timæus of Plato) did make the world. That,one turnyng, should rule motions most vnlike in slownes, and swiftnes.* But a greater cause of meruayling we haue by *Claudianus* report hereof. Who affirmeth this *Archimedes worke*,to haue ben of Glasse. And discourseth of it more at large:which I omit. The Doue of wood, which the *Mathematicien Archytas* did make to flye,is by *Agellius* spoken of.Of *Dædalus* straunge Images, *Plato* reporteth.*Homere* of *Vulcans Selfmouers,(* by secret wheles)leaueth in writyng. *Aristotle*,in hys *Politikes*,of both, maketh mention. Meruaylous was the workemanshyp,of late dayes,performed by good skill of *Trochilike. &c*. For in Noremberge, A flye of Iern,beyng let out of the Artificers hand,did(as it were)fly about by the gestes,at the table,and at length,as though it were weary, retourne to his masters hand agayne. Moreouer, an Artificiall Egle, was ordred, to fly out of the same Towne,a mighty way,and that a loft in the Ayre, toward the Emperour comming thether:and followed hym,beyng come to the gate of the towne.* Thus, you see, what, Arte Mathematicall can performe,when Skill, will, Industry, and Hability, are duely applyed to profe.

*Anno. 1551

De his quæ Mundo mirabiliter eueniunt. cap.8.

Tusc. 1.

☞ *

A Digression. Apologeticall.

AND for these,and such like marueilous Actes and Feates,Naturally,Mathematically,and Mechanically, wrought and contriued : ought any honest Student, and Modest Christian Philosopher,be counted,& called a Coniurer ? Shall the folly of Idiotes, and the Mallice of the Scornfull, so much preuaile, that He, who seeketh no worldly gaine or glory at their handes : But onely,of God,the threasor of heauenly wisedome,& knowledge of pure veritie : Shall he (I say) in the meane

space,

ſpace, be robbed and ſpoiled of his honeſt name and fame? He that ſeketh (by S. Paules aduertiſement) in the Creatures Properties, and wonderfull vertues, to finde iuſte cauſe, to glorifie the Æternall, and Almightie Creator by: Shall that man, be (in hugger mugger) condemned, as a Companion of the Helhoundes, and a Caller, and Coniurer of wicked and damned Spirites? He that bewaileth his great want of time, ſufficient (to his contentation) for learning of Godly wiſdome, and Godly Verities in: and onely therin ſetteth all his delight: Will that mã leeſe and abuſe his time, in dealing with the Chiefe enemie of Chriſt our Redemer: the deadly foe of all mankinde: the ſubtile and impudent peruerter of Godly Veritie: the Hypocriticall Crocodile: the Enuious Baſiliſke, continually deſirous, in the twinke of an eye, to deſtroy all Mankinde, both in Body and Soule, æternally? Surely (for my part, ſomewhat to ſay herein) I haue not learned to make ſo brutiſh, and ſo wicked a Bargaine. Should I, for my xx. or xxv. yeares Studie: for two or three thouſand Markes ſpending: ſeuen or eight thouſand Miles going and trauailing, onely for good learninges ſake: And that, in all maner of wethers: in all maner of waies and paſſages: both early and late: in daunger of violence by man: in daunger of deſtruction by wilde beaſtes: in hunger: in thirſt: in perilous heates by day, with toyle on foote: in daungerous dampes of colde, by night, almoſt bereuing life: (as God knoweth): with lodginges, oft times, to ſmall eaſe: and ſomtime to leſſe ſecuritie. And for much more (then all this) done & ſuffred, for Learning and attaining of Wiſedome: Should I (I pray you) for all this, no otherwiſe, nor more warily: or (by Gods mercifulnes) no more luckily, haue fiſhed, with ſo large, and coſtly, a Nette, ſo long time in drawing (and that with the helpe and aduiſe of Lady Philoſophie, & Queene Theologie): but at length, to haue catched, and drawen vp,* a Frog? Nay, a Deuill? For, ſo, doth the Common peuiſh Pratler Imagine and Iangle: And, ſo, doth the Malicious ſkorner, ſecretly wiſhe, & brauely and boldly face down, behinde my backe. Ah, what a miſerable thing, is this kinde of Men? How great is the blindnes & boldnes, of the Multitude, in thinges aboue their Capacitie? What a Land: what a People: what Maners: what Times are theſe? Are they become Deuils, them ſelues: and, by falſe witneſſe bearing againſt their Neighbour, would they alſo, become Murderers? Doth God, ſo long geue them reſpite, to reclaime them ſelues in, from this horrible ſlaundering of the giltleſſe: contrary to their owne Conſciences: and yet will they not ceaſe? Doth the Innocent, forbeare the calling of them, Iuridically to aunſwere him, according to the rigour of the Lawes: and will they deſpiſe his Charitable pacience? As they, againſt him, by name, do forge, fable, rage, and raiſe ſlaunder, by Worde & Print: Will they prouoke him, by worde and Print, likewiſe, to Note their Names to the World: with their particular deuiſes, fables, beaſtly Imaginations, and vnchriſtenlike ſlaunders? Well: Well. O (you ſuch) my vnkinde Countrey men. O vnnaturall Countrey men. O vnthankfull Countrey men. O Brainſicke, Raſhe, Spitefull, and Diſdainfull Countrey men. Why oppreſſe you me, thus violently, with your ſlaundering of me: Contrary to Veritie: and contrary to your owne Conſciences? And I, to this hower, neither by worde, deede, or thought, haue bene, any way, hurtfull, damageable, or iniurious to you, or yours? Haue I, ſo long, ſo dearly, ſo farre, ſo carefully, ſo painfully, ſo daungerouſly ſought & trauailed for the learning of Wiſedome, & atteyning of Vertue: And in the end (in your iudgemẽt) am I become, worſe, then when I begã? Worſe, thẽ a Mad man? A dangerous Member in the Common Wealth: and no Member of the Church of Chriſt? Call you this, to be Learned? Call you this, to be a Philoſopher? and a louer of Wiſedome? To forſake the ſtraight heauenly way: and to wallow in the broad way of

** A prouerb. Fayre fiſht, and caught a Frog.*

damnation? To forſake the light of heauenly Wiſedome: and to lurke in the dungeon of the Prince of darkeneſſe? To forſake the Veritie of God,& his Creatures: and to fawne vpon the Impudent,Craftie,Obſtinate Lier, and continuall diſgracer of Gods Veritie, to the vttermoſt of his power? To forſake the Life & Bliſſe Æternall: and to cleaue vnto the Author of Death euerlaſting? that Murderous Tyrant, moſt gredily awaiting the Pray of Mans Soule? Well: I thanke God and our Lorde Ieſus Chriſt, for the Comfort which I haue by the Examples of other men, before my time: To whom,neither in godlines of life, nor in perfection of learning, I am worthy to be compared: and yet, they ſuſtained the very like Iniuries, that I do: or rather,greater. Pacient *Socrates*, his *Apologie* will teſtifie: *Apuleius* his *Apologies*, will declare the Brutiſhneſſe of the Multitude. *Ioannes Picus*, Earle of Mirandula, his *Apologie* will teach you, of the Raging ſlaunder of the Malicious Ignorant againſt him. *Ioannes Trithemius*, his *Apologie* will ſpecifie, how he had occaſion to make publike Proteſtation: as well by reaſon of the Rude Simple: as alſo,in reſpect of ſuch,as were counted to be of the wiſeſt ſort of men. Many could I recite: But I deferre the preciſe and determined handling of this matter: being loth to detect the Folly & Mallice of my Natiue Countrey men.* Who, ſo hardly, can diſgeſt or like any extraordinary courſe of Philoſophicall Studies: not falling within the Cumpaſſe of their Capacitie: or where they are not made priuie of the true and ſecrete cauſe, of ſuch wonderfull Philoſophicall Feates. These men, are of fower ſortes, chiefly. The firſt, I may name, *Vaine pratling buſie bodies*: The ſecond, *Fond Frendes*: The third, *Imperfectly zelous*: and the fourth, *Malicious Ignorant*. To eche of theſe (briefly,and in charitie) I will ſay a word or two, and ſo returne to my Præface.

1\. *Vaine pratling buſie bodies*, vſe your idle aſſemblies,and conferences, otherwiſe, then in talke of matter, either aboue your Capacities, for hardneſſe: or contrary to your Conſciences, in Veritie.

2\. *Fonde Frendes*, leaue of,ſo to commend your vnacquainted frend,vpon blinde affection: As, becauſe he knoweth more,then the common Student: that, therfore, he muſt needes be ſkilfull, and a doer, in ſuch matter and maner, as you terme *Coniuring*. Weening,thereby, you aduaunce his fame: and that you make other men, great marueilers of your hap, to haue ſuch a learned frend. Ceaſe to aſcribe Impietie, where you pretend Amitie. For,if your tounges were true, then were that your frend, *Vntrue*, both to God,and his Soueraigne. Such *Frendes* and *Fondlinges*, I ſhake of, and renounce you: Shake you of, your Folly.

3\. *Imperfectly zelous*,to you, do I ſay: that (perhaps) well, do you Meane: But farre you miſſe the Marke: If a Lambe you will kill, to feede the flocke with his bloud. Sheepe, with Lambes bloud, haue no naturall ſuſtenaunce: No more, is Chriſtes flocke, with horrible ſlaunders, duely ædified. Nor your faire pretenſe, by ſuch raſhe ragged Rhetorike,any whit,well graced. But ſuch,as ſo vſe me,will finde a fowle Cracke in their Credite. Speake that you know: And know, as you ought: Know not,by Heare ſay, when life lieth in daunger. Search to the quicke,& let Charitie be your guide.

4\. *Malicious Ignorant*, what ſhall I ſay to thee? *Prohibe linguam tuam a malo. A detractione parcite lingua*. *Cauſe thy toung to refraine frõ euill. Refraine your toung from ſlaunder*. Though your tounges be ſharpned, Serpent like, & Adders poyſon lye in your lippes: yet take heede,and thinke,betimes, with your ſelfe, *Vir linguoſus non ſtabilietur in terra. Virum violentum venabitur malum, donec præcipitetur.* For,ſure I am, *Quia faciet Dominus Iudicium afflicti: & vindictam pauperum.* (*Pſal.* 140.)

Thus, I require you, my aſſured frendes, and Countrey men (you Mathematiciens, Mechaniciens,and Philoſophers, Charitable and diſcrete) to deale in my behalfe, behalf,with the light & vntrue tounged, my enuious Aduersaries,or Fond frends. And farther, I would wishe, that at leysor, you would consider,how *Basilius Magnus,* layeth *Moses* and *Daniel,* before the eyes of those, which count all such Studies Philosophicall (as mine hath bene) to be vngodly, or vnprofitable. Waye well *S.Stephen* his witnesse of *Moses*. *Eruditus est Moses omni Sapientia Ægyptiorũ: & erat potens in verbis & operibus suis*. *Moses was instructed in all maner of wisedome of the Ægyptians: and he was of power both in his wordes, and workes.* You see this Philosophicall Power & Wisedome,which *Moses* had,to be nothing misliked of the Holy Ghost. Yet *Plinius* hath recorded,*Moses* to be a wicked *Magicien*. And that (of force) must be, either for this Philosophicall wisedome,learned, before his calling to the leading of the Children of *Israel:* or for those his wonders,wrought before King *Pharao,* after he had the conducting of the *Israelites*. As concerning the first,you perceaue, how *S.Stephen,* at his Martyrdome (being full of the Holy Ghost) in his Recapitulation of the olde Testament, hath made mention of *Moses* Philosophie: with good liking of it: And *Basilius Magnus* also, auoucheth it, to haue bene to *Moses* profitable (and therefore, I say, to the Church of God, necessary). But as cõcerning *Moses* wonders,done before King *Pharao*: God, him selfe, sayd: *Vide vt omnia ostenta, quæ posui in manu tua, facias coram Pharaone. See that thou do all those wonders before Pharao, which I haue put in thy hand.* Thus, you euidently perceaue,how rashly,*Plinius* hath slaundered *Moses*, of vayne fraudulent *Magike*, saying: *Est & alia Magices Factio, a Mose, Iamne,& Iotape, Iudæis pendens: sed multis millibus annorum post Zoroastrem.&c.* Let all such, therefore, who, in Iudgement and Skill of Philosophie, are farre Inferior to *Plinie*, take good heede, least they ouershoote them selues rashly, in Iudging of *Philosophers straunge Actes*· and the Meanes,how they are done. But, much more,ought they to beware of forging, deuising, and imagining monstrous feates, and wonderfull workes, when and where, no such were done: no, not any sparke or likelihode,of such,as they,without all shame, do report. And (to conclude) most of all, let them be ashamed of Man, and afraide of the dreadfull and Iuste Iudge: both Folishly or Maliciously to deuise: and then,deuilishly to father their new fond Monsters on me: Innocent, in hand and hart: for trespacing either against the lawe of God, or Man, in any my Studies or Exercises, Philosophicall, or Mathematicall: As in due time, I hope, will be more manifest.

Act.7.C.

Lib.30. Cap.1.

1. ,,

,, ☞

2.

3.

NOw end I,with **Archemastrie.** Which name, is not so new,as this Arte is rare. For an other Arte,vnder this,a degree (for skill and power) hath bene indued with this English name before. And yet,this,may serue for our purpose, sufficiently,at this present. This Arte, teacheth to bryng to actuall experience sensible,all worthy conclusions by all the Artes Mathematicall purposed, & by true Naturall Philosophie concluded: & both addeth to them a farder scope,in the termes of the same Artes, & also by hys propre Method,and in peculier termes, procedeth, with helpe of the foresayd Artes, to the performance of complet Experiẽces,which of no particular Art, are hable (Formally) to be challenged. If you remember,how we considered *Architecture,* in respect of all common handworkes: some light may you haue,therby,to vnderstand the Souerainty and propertie of this Science. *Science* I may call it,rather, then an Arte: for the excellency and Mastershyp it hath, ouer so many, and so mighty Artes and

 Sciences.

Sciences. And bycause it procedeth by *Experiences*, and searcheth forth the causes of Conclusions, by *Experiences*: and also putteth the Conclusions them selues, in *Experience*, it is named of some, *Scientia Experimentalis*. The *Experimentall Science*. *Nicolaus Cusanus* termeth it so, in hys *Experimentes Statikall*, And an other *Philosopher*, of this land Natiue (the floure of whose worthy fame, can neuer dye nor wither) did write therof largely, at the request of *Clement the sixt*. The Arte carrieth with it, a wonderfull Credit: By reason, it certefieth, sensibly, fully, and completely to the vtmost power of Nature, and Arte. This Arte, certifieth by *Experience* complete and absolute: and other Artes, with their Argumentes, and Demonstrations, persuade: and in wordes, proue very well their Conclusions. * But wordes, and Argumentes, are no sensible certifying: nor the full and finall frute of Sciences practisable. And though some Artes, haue in them, *Experiences*, yet they are not complete, and brought to the vttermost, they may be stretched vnto, and applyed sensibly. As for example: the Naturall Philosopher disputeth and maketh goodly shew of reason: And the Astronomer, and the Opticall Mechanicien, put some thynges in *Experience*: but neither, all, that they may: nor yet sufficiently, and to the vtmost, those, which they do, There, then, the *Archemaster* steppeth in, and leadeth forth on, the *Experiences*, by order of his doctrine *Experimentall*, to the chief and finall power of Naturall and Mathematicall Artes. Of two or three men, in whom, this Description of *Archemastry* was *Experimentally*, verified, I haue read and hard: and good record, is of their such perfection. So that, this Art, is no fantasticall Imagination: as some Sophister, might, *Cum suis Insolubilibus*, make a florish: and dassell your Imagination: and dash your honest desire and Courage, from beleuing these thinges, so vnheard of, so meruaylous, & of such Importance. Well: as you will. I haue forewarned you. I haue done the part of a frende: I haue discharged my Duety toward God: for my small Talent, at hys most mercyfull handes receiued. To this Science, doth the *Science Alnirangiat*, great Seruice. Muse nothyng of this name. I chaunge not the name, so vsed, and in Print published by other: beyng a name, propre to the Science. Vnder this, commeth *Ars Sintrillia*, by *Artephius*, briefly written. But the chief Science, of the Archemaster, (in this world) as yet knowen, is an other (as it were) OPTICAL Science: wherof, the name shall be told (God willyng) when I shall haue some, (more iust) occasion, therof, to Discourse.

R. B.

☞

Here, I must end, thus abruptly (Gentle frende, and vnfayned louer of honest and necessary verities.) For, they, who haue (for your sake, and vertues cause) requested me, (an old forworne Mathematicien) to take pen in hand: (through the confidence they reposed in my long experience: and tryed sincerity) for the declaryng and reportyng somewhat, of the frute and commodity, by the Artes Mathematicall, to be atteyned vnto: euen they, Sore agaynst their willes, are forced, for sundry causes, to satiffie the workemans request, in endyng forthwith: He, so feareth this, so new an attempt, & so costly: And in matter so slenderly (hetherto) among the common Sorte of Studentes, considered or estemed.

And where I was willed, somewhat to alledge, why, in our vulgare Speche, this part of the Principall Science of *Geometrie*, called *Euclides Geometricall Elementes*, is published, to your handlyng: being vnlatined people, and not Vniuersitie Scholers: Verily, I thinke it nedelesse.

1. For, the Honour, and Estimation of the Vniuersities, and Graduates, is, hereby, nothing diminished. Seing, from, and by their Nurse Children, you receaue all this Benefite: how great soeuer it be.

Neither

Neither are their Studies, hereby, any whit hindred. No more, then the Italian *Vniuersities*, as *Academia Bononiensis, Ferrariensis, Florentina, Mediolanensis, Patauina, Papiensis, Perusina, Pisana, Romana, Senensis*, or any one of them, finde them selues, any deale, disgraced, or their Studies any thing hindred, by *Frater Lucas de Burgo*, or by *Nicolaus Tartalea*, who in vulgar Italian language, haue published, not onely *Euclides Geometrie*, but of *Archimedes* somewhat: and in Arithmetike and Practicall Geometrie, very large volumes, all in their vulgar speche. Nor in Germany haue the famous *Vniuersities*, any thing bene discontent with *Albertus Durerus*, his Geometricall Institutions in Dutch: or with *Gulielmus Xylander*, his learned translation of the first sixe bookes of *Euclide*, out of the Greke into the high Dutch. Nor with *Gualterus H. Riffius*, his Geómetricall Volume: very diligently translated into the high Dutch tounge, and published. Nor yet the *Vniuersities* of Spaine, or Portugall, thinke their reputation to be decayed: or suppose any their Studies to be hindred by the Excellent *P. Nonnius*, his Mathematicall workes, in vulgare speche by him put forth. Haue you not, likewise, in the French tounge, the whole Mathematicall Quadriuie? and yet neither Paris, Orleance, or any of the other Vniuersities of Fraunce, at any time, with the Translaters, or Publishers offended: or any mans Studie thereby hindred? 2.

And surely, the Common and Vulgar Scholer (much more, the Gramarian) before his comming to the *Vniuersitie*, shall (or may) be, now (according to *Plato* his Counsell) sufficiently instructed in *Arithmetike* and *Geometrie*, for the better and easier learning of all maner of *Philosophie, Academicall*, or *Peripateticall*. And by that meanes, goe more cherefully, more skilfully, and spedily forwarde, in his Studies, there to be learned. And, so, in lesse time, profite more, then (otherwise) he should, or could do. 3.

Also many good and pregnant Englishe wittes, of young Gentlemen, and of other, who neuer intend to meddle with the profound search and Studie of Philosophie (in the *Vniuersities* to be learned) may neuerthelesse, now, with more ease and libertie, haue good occasion, vertuously to occupie the sharpnesse of their wittes: where, els (perchance) otherwise, they would in fond exercises, spend (or rather leese) their time: neither seruing God: nor furdering the Weale, common or priuate. 4.

And great Comfort, with good hope, may the *Vniuersities* haue, by reason of this *Englishe* Geometrie, and Mathematicall Præface, that they (hereafter) shall be the more regarded, esteemed, and resorted vnto. For, when it shall be knowen and reported, that of the *Mathematicall Sciences* onely, such great Commodities are ensuing (as I haue specified): and that in dede, some of you vnlatined Studentes, can be good witnesse, of such rare fruite by you enioyed (thereby): as either, before this, was not heard of: or els, not so fully credited: Well, may all men " coniecture, that farre greater ayde, and better furniture, to winne to the Perfection " of all Philosophie, may in the Vniuersities be had: being the Storehouses & Threasory of all Sciences, and all Artes, necessary for the best, and most noble State of " Common Wealthes. " 5.

Vniuersities.

Besides this, how many a Common Artificer, is there, in these Realmes of England and Ireland, that dealeth with Numbers, Rule, & Cumpasse: Who, with their owne Skill and experience, already had, will be hable (by these good helpes and informations) to finde out, and deuise, new workes, straunge Engines, and Instrumentes: for sundry purposes in the Common Wealth? or for priuate pleasure? and for the better maintayning of their owne estate? I will not (therefore) 6.

 fight

fight against myne owne shadowe. For, no man (I am sure) will open his mouth against this Enterprise. No mã (I say) who either hath Charitie toward his brother (and would be glad of his furtherance in vertuous knowledge): or that hath any care & zeale for the bettering of the Cõmon state of this Realme. Neither any, that make accompt, what the wiser sort of men (Sage and Stayed) do thinke of them. To none (therefore) will I make any *Apologie*, for a vertuous acte doing: and for cõmending, or setting forth, Profitable Artes to English men, in the English toung.
„ But, vnto God our Creator, let vs all be thankefull: for that, *As he, of his Goodnes, by his Powre, and in his wisedome, hath Created all thynges, in Number, Waight, and Measure*: So, to vs, of hys great Mercy, he hath reuealed Meanes, whereby, to atteyne the sufficient and necessary knowledge of the foresayd hys three principall Instrumentes: Which Meanes, I haue abundantly proued vnto you, to be the *Sciences* and *Artes Mathematicall*.

And though I haue ben pinched with straightnes of tyme: that, no way, I could so pen downe the matter (in my Mynde) as I determined: hopyng of conuenient laysure: Yet, if vertuous zeale, and honest Intent prouoke and bryng you to the readyng and examinyng of this Compendious treatise, I do not doute, but, as the veritie therof (accordyng to our purpose) will be euident vnto you: So the pith and force therof, will persuade you: and the wonderfull frute therof, highly pleasure you. And that you may the easier perceiue, and better remember, the principall pointes, whereof my Preface treateth, I will giue you the Groundplatt of my whole discourse, in a Table annexed: from the first to the last, somewhat Methodically contriued.

The Ground platt of this Præface in a Table.

If Hast, hath caused my poore pen, any where, to stumble: You will, (I am sure) in part of recompence, (for my earnest and sincere good will to pleasure you), Consider the rockish huge mountaines, and the perilous vnbeaten wayes, which (both night and day, for the while) it hath toyled and labored through, to bryng you this good Newes, and Comfortable profe, of Vertues frute.

So, I Commit you vnto Gods Mercyfull direction, for the rest: hartely besechyng hym, to prosper your Studyes, and honest Intentes: to his Glory, & the Commodity of our Countrey. *Amen.*

Written at my poore House At Mortlake.

Anno. 1570. February. 9.